Periodic Table of the Elements with the Gmelin System Numbers

1	2	3	4	5	6	7	8	9	10	11	12	13	14	15	16	17	18
1 H 2																1 H 2	2 He I
3 Li 20	4 Be 26											5 B 13	6 C 14	7 N 4	8 O 3	9 F 5	10 Ne I
11 Na 21	12 Mg 27											13 Al 35	14 Si 15	15 P 16	16 S 9	17 Cl 6	18 Ar I
19 * K 22	20 Ca 28	21 Sc 39	22 Ti 41	23 V 48	24 Cr 52	25 Mn 56	26 Fe 59	27 Co 58	28 Ni 57	29 Cu 60	30 Zn 32	31 Ga 36	32 Ge 45	33 As 17	34 Se 10	35 Br 7	36 Kr I
37 Rb 24	38 Sr 29	39 Y 39	40 Zr 42	41 Nb 49	42 Mo 53	43 Tc 69	44 Ru 63	45 Rh 64	46 Pd 65	47 Ag 61	48 Cd 33	49 In 37	50 Sn 46	51 Sb 18	52 Te II	53 I 8	54 Xe I
55 Cs 25	56 Ba 30	57** La 39	72 Hf 43	73 Ta 50	74 W 54	75 Re 70	76 Os 70	77 Ir 67	78 Pt 68	79 Au 62	80 Hg 34	81 Tl 38	82 Pb 47	83 Bi 19	84 Po 12	85 At I	86 Rn I
87 Fr 31	88 Ra 31	89*** Ac 40	104 71	105 71													

**Lanthanides 39	58 Ce 39	59 Pr	60 Nd	61 Pm	62 Sm	63 Eu	64 Gd	65 Tb	66 Dy	67 Ho	68 Er	69 Tm	70 Yb	71 Lu
***Actinides	90 Th 44	91 Pa 51	92 U 55	93 Np 71	94 Pu 71	95 Am 71	96 Cm 71	97 Bk 71	98 Cf 71	99 Es 71	100 Fm 71	101 Md 71	102 No 71	103 Lr 71

*
NH₄ 23 → NH_4 23

Gmelin Handbook of Inorganic Chemistry

8th Edition

"Gmelin Handbook Volumes on Radium and Actinides"
see pages VIII to X

Gmelin Handbook
of Inorganic Chemistry

8th Edition

Gmelin Handbuch der Anorganischen Chemie

Achte, völlig neu bearbeitete Auflage

Prepared
and issued by

Gmelin-Institut für Anorganische Chemie
der Max-Planck-Gesellschaft
zur Förderung der Wissenschaften

Director: Ekkehard Fluck

Founded by

Leopold Gmelin

8th Edition

8th Edition begun under the auspices of the
Deutsche Chemische Gesellschaft by R. J. Meyer

Continued by

E. H. E. Pietsch and A. Kotowski, and by
Margot Becke-Goehring

Springer-Verlag Berlin Heidelberg GmbH 1985

Gmelin Handbook of Inorganic Chemistry

8th Edition

Th
Thorium

Supplement Volume C 5

Compounds with S, Se, Te, and B

With 58 illustrations

AUTHORS

David Brown, Chemistry Division
A.E.R.E., Harwell, Oxon, England

Horst Wedemeyer, Institut für Material-
und Festkörperforschung,
Kernforschungszentrum Karlsruhe

CHIEF EDITORS

Karl-Christian Buschbeck, Gmelin-Institut, Frankfurt am Main

Cornelius Keller, Supervising Scientific Coordinator
for the Thorium Supplement Volumes,
Schule für Kerntechnik, Kernforschungszentrum Karlsruhe

System Number 44

Springer-Verlag Berlin Heidelberg GmbH 1985

LITERATURE CLOSING DATE: 1984
IN SOME CASES MORE RECENT DATA HAVE BEEN CONSIDERED

Library of Congress Catalog Card Number: Agr 25-1383

ISBN 978-3-662-06344-6   ISBN 978-3-662-06342-2 (eBook)
DOI 10.1007/978-3-662-06342-2

© by Springer-Verlag Berlin Heidelberg 1985
Originally published by Springer-Verlag Berlin Heidelberg New York Tokyo in 1985
Softcover reprint of the hardcover 8th edition 1985

Preface

The present volume, Thorium C5, deals with the compounds of thorium and sulfur, selenium, tellurium, and boron, as well as with oxoacid compounds of the three chalcogen elements. Thorium borates have already been treated in Thorium C2. In contrast to the corresponding compounds of uranium the thorium sulfides, etc., do not show any nuclear or other technological application; they are only of academic interest, despite some very interesting electronic properties, especially of the 1:1 compounds. The thorium-sulfur and the thorium-boron systems in particular were studied in detail, so that we have a clear picture of them, whereas there are still a lot of open questions in the systems Th-Se and Th-Te — not very different from other metal chalcogenide systems. Thorium sulfates are of some technological importance because they are formed in solution during recovery of thorium from monazite by sulfuric acid leaching. The very detailed and critical treatment of the chemical and physical properties of the compounds discussed also enables us to find gaps still remaining in our knowledge and thus to initiate new research in this field.

I want to thank the two authors, Dr. Horst Wedemeyer (Karlsruhe) and Dr. David Brown (Harwell), for their excellent contributions, the "Literaturabteilung" of the Karlsruhe Nuclear Research Center for its help in providing reports and other documents difficult to procure, as well as the staff of the Gmelin-Institute, especially to Dr. K.-C. Buschbeck as the responsible editor and to the Director Prof. Dr. E. Fluck for the excellent cooperation.

Karlsruhe
Christmas 1985

Cornelius Keller

Volumes published on "Radium and Actinides"

Ac   Actinium

Main Volume — 1942

Suppl. Vol. 1: Element and Compounds — 1981

Np, Pu, ... Transuranium Elements

Main Volume

Part A: The Elements

A 1, I   History, Occurrence, Properties of Atomic Nuclei — 1973
A 1, II Nuclides: Manufacture, Recovery, Enrichment — 1974
A 2   General Properties, Uses, Storage, Biology — 1973

Part B: The Metals

B 1   Metals — 1976
B 2   Binary Alloy Systems 1 — 1976
B 3   Binary Alloy Systems 2 — 1977

Part C: The Compounds

C   Compounds — 1972

Part D: Chemistry in Solution

D 1   Aqueous Solutions. Coordination Chemistry — 1975
D 2   Extraction, Ion Exchange. Molten Salts — 1975

Index — 1979

Pa   Protactinium

Main Volume — 1942

Suppl. Vol. 1: Element — 1977

Suppl. Vol. 2: Metal. Alloys. Compounds. Chemistry in Solution — 1977

Ra   Radium

Main Volume — 1928

Suppl. Vol. 1: History. Cosmochemistry. Geochemistry — 1977

Suppl. Vol. 2: Element. Compounds — 1977

Th Thorium

Main Volume — 1955

Suppl. Vol. Part C: The Compounds

C1 Compounds with Noble Gases, Hydrogen, Oxygen — 1978
C2 Ternary and Polynary Oxides — 1976
C5 Compounds with S, Se, Te, B (**present volume**) — 1986

Suppl. Vol. Part D: Chemistry in Solution

D2 Solvent Extraction — 1985

Suppl. Vol. Part E: Coordination Compounds

E Coordination Compounds — 1985

U Uranium

Main Volume — 1936

Suppl. Vol. Part A: The Element

A1 Uranium Deposits — 1979
A2 Isotopes — 1980
A3 Technology. Uses — 1981
A4 Irradiated Fuel. Reprocessing — 1982
A5 Spectra — 1982
A6 General Properties. Criticality — 1983
A7 Analysis. Biology — 1982

Suppl. Vol. Part B: The Alloys (in preparation)

Suppl. Vol. Part C: The Compounds

C1 Compounds with Noble Gases and Hydrogen. Uranium-Oxygen System — 1977
C2 Oxides U_3O_8 and UO_3. Hydroxides, Oxide Hydrates, and Peroxides — 1978
C3 Ternary and Polynary Oxides — 1975
C4 UO_2, Preparation and Crystallographic Properties — 1984
C5 UO_2, Physical Properties. Electrochemical Behavior — 1986
C6 UO_2, Chemical Properties (in preparation)
C7 Compounds with Nitrogen — 1981
C8 Compounds with Fluorine — 1980
C9 Compounds with Chlorine, Bromine, and Iodine — 1979
C10 Compounds with Sulfur — 1984
C11 Compounds with Selenium, Tellurium, and Boron — 1981
C12 Carbides (in preparation)
C13 Carbonates, Cyanides, Thiocyanates, Alkoxides, Carboxylates. Compounds with Silicon — 1983
C14 Compounds with Phosphorus, Arsenic, Antimony, Bismuth, and Germanium — 1981

X

Suppl. Vol. Part D: Chemistry in Solution

D1	Properties of the Ions. Molten Salts	— 1984
D2	Solvent Extraction	— 1982
D3	Anion Exchange	— 1982
D4	Cation Exchange and Chromatography	— 1983

Suppl. Vol. Part E: Coordination Compounds

E1	Coordination Compounds 1	— 1979
E2	Coordination Compounds 2 (including Organouranium Compounds)	— 1980

Table of Contents

Page

10   Compounds of Thorium and Sulfur 1

10.1   Binary Thorium Sulfides 1

10.1.1   The Th-S System 1

10.1.2   Thorium Monosulfide, ThS 2

Formation and Preparation 2
   Synthesis from the Elements 2
   Reaction of Thorium Hydride with H_2S 3
   Reduction of ThS_2 3
   Reduction of ThOS 3
   Reaction of Thorium with ZnS 3
   Fused Salt Electrolysis 4
   Sintering Behavior 4
   Purification of ThS 5
   Single Crystals 5
   Enthalpy, Entropy, and Gibbs Free Energy of Formation 5
Crystallographic Properties, Bonding, and Lattice Dynamics 5
Mechanical Properties 9
   Density 9
   Elasticity, Hardness, and Strength 9
Thermal Properties 9
   Thermal Expansion 9
   Vaporization 9
   Melting Point 10
   Heat Capacity and Thermodynamic Functions 10
   Thermal Conductivity 14
Electrical Properties 17
   Electronic Structure 17
   Electric Resistivity 18
   Thermoelectric Power 19
   Hall Coefficient 20
   Thermionic Emission of Electrons 20
Magnetic Properties 20
Optical Properties 20
Chemical Reactions 20
   On Heating 20
   With Elements 21
   With Compounds 21

10.1.3   Dithorium Trisulfide, Th_2S_3 22

Formation and Preparation 22
   Reduction of ThS_2 22
   Reaction of Thorium Hydride with H_2S 22
   Decomposition of ThS_2 22
   Synthesis from the Elements 22

Page

Densification of Th_2S_3 22
Enthalpy, Entropy, and Gibbs Free Energy of Formation 23
Crystallographic Properties 23
Mechanical Properties 24
Thermal Properties 24
Electrical Properties 26
Magnetic Properties 26
Chemical Reactions 26
On Heating 26
With Elements 26
With Compounds 27

10.1.4   Heptathorium Dodecasulfide, Th_7S_{12} 27

Formation and Preparation 27
Decomposition of ThS_2 27
Reaction of Thorium or Thorium Hydride with H_2S 27
Attempted Synthesis from the Elements 28
Other Methods 28
Densification 28
Enthalpy, Entropy, and Gibbs Free Energy of Formation 28
Crystallographic Properties 29
Mechanical Properties 30
Thermal Properties 30
Electrical Properties 30
Magnetic Properties 30
Chemical Reactions 30
With Elements 30
With Compounds 31

10.1.5   Thorium Disulfide, ThS_2 32

Formation and Preparation 32
Synthesis from the Elements 32
Reaction of Thorium or Thorium Hydride with H_2S 32
Reaction of ThO_2 with H_2S in the Presence of Carbon 33
Other Methods 33
Densification 35
Single Crystals 35
Enthalpy, Entropy, and Gibbs Free Energy of Formation 35
Crystallographic Properties 35
Thermal Properties 36
Vaporization. Melting Point 36
Heat Capacity and Thermodynamic Functions 36
Electrical Properties 41
Magnetic Properties 41
Chemical Reactions 41
On Heating 41
With Elements 41
With Compounds 42

		Page
10.1.6 Dithorium Pentasulfide, Th_2S_5		43
Formation and Preparation		43
Crystallographic Properties		43
Thermal Properties		44
Electrical Properties		45
Chemical Behavior		45
10.2 Compounds of Thorium with Sulfur and Oxygen		48
10.2.1 Thorium Oxide Sulfide, ThOS		48
Formation and Preparation		48
Reaction of Thorium Oxide with H_2S		49
Reaction of Thorium Oxide with Sulfur		49
Reaction of Thorium Oxide with ThS_2		49
Single Crystals		49
Enthalpy of Formation		50
Crystallographic Properties		50
Heat Capacity and Entropy		51
Chemical Reactions		52
With Elements		52
With Compounds		52
10.3 Compounds of Thorium with Sulfur and Nitrogen		53
10.3.1 Dithorium Dinitride Sulfide, Th_2N_2S		53
Preparation		53
Crystallographic Properties		53
10.4 Compounds of Thorium with Rare Earth Elements and Sulfur		53
10.4.1 Phase Relationships		53
10.4.2 LnS-ThS Solid Solutions		54
Chemical Reactions		54
10.4.3 $Y_{0.84}Th_{0.16}S_{1.3}$		55
10.4.4 $Ln_2^{III}ThS_5$ Compounds		55
10.4.5 $Ln_4^{III}Th_5S_{16}$ Compounds		56
10.4.6 $CeThS_2$		56
10.5 Thorium Sulfite Compounds		57
10.5.1 Introduction		57
10.5.2 Thorium(IV) Sulfites		57
Preparation		57
Properties		58

Page

10.5.3   Thorium(IV) Basic Sulfites   59

Preparation   59
Properties   59

10.5.4   Thorium(IV) Mixed Acid Sulfites   60

10.5.5   Thorium(IV) Sulfito Complexes   60

Preparation   60
Properties   61

10.5.6   Thorium(IV) Hydroxo Sulfito Complexes   62

10.5.7   Thorium(IV) Mixed Acid Sulfito Complexes   62

10.6   Thorium Sulfate Compounds   63

10.6.1   Introduction   63

10.6.2   Thorium Bis(sulfate), $Th(SO_4)_2$ 63

Preparation   63
Physical Properties   63
Chemical Properties   66

10.6.3   Thorium Bis(sulfate) Hydrates   67

Preparation   67
Physical Properties   68
Chemical Properties   69

10.6.4   Thorium Oxide Sulfate   75

10.6.5   Thorium Peroxide Sulfate Hydrates   76

10.6.6   Thorium Hydroxide Sulfates   76

10.6.7   Thorium(IV) Mixed Acid Sulfates   77

Thorium(IV) Oxydiacetate Sulfate Hydrate   77

10.6.8   Thorium(IV) Sulfato Complexes   79

Preparation   79
Physical Properties   81
Chemical Properties   82

10.6.9   Thorium(IV) Mixed Acid Sulfato Complexes   87

10.7   Thorium(IV) Fluorosulfate   87

11   Compounds of Thorium and Selenium   88

11.1   Binary Thorium Selenides   88

11.1.1   The Th-Se System   88

Page

11.1.2   Thorium Monoselenide, $ThSe_2$ 89

　Formation and Preparation   89
　　Synthesis from the Elements   89
　　Reduction of $ThSe_2$ 90
　　Reaction of Thorium with ZnSe   90
　　Enthalpy and Gibbs Free Energy of Formation   90
　Crystallographic Properties, Bonding, and Lattice Dynamics   90
　Thermal Properties   91
　Electrical Properties   93
　Magnetic Properties   93
　Chemical Behavior   93

11.1.3   Dithorium Triselenide, Th_2S_3   94

　Formation and Preparation   94
　　Synthesis from the Elements   94
　　Enthalpy and Gibbs Free Energy of Formation   94
　Crystallographic Properties   94
　Thermal Properties   94
　Electrical Properties   95
　Chemical Behavior   95

11.1.4   Heptathorium Dodecaselenide, Th_7Se_{12}   95

　Formation and Preparation   95
　　Synthesis from the Elements   95
　　Enthalpy and Gibbs Free Energy of Formation   95
　Crystallographic Properties   95
　Thermal Properties   96
　Electrical Properties   96
　Chemical Behavior   96

11.1.5   Thorium Diselenide, $ThSe_2$ 96

　Formation and Preparation   96
　　Synthesis from the Elements   97
　　Enthalpy and Gibbs Free Energy of Formation   97
　Crystallographic Properties   97
　Thermal Properties   97
　Electrical Properties   99
　Chemical Behavior   99

11.1.6   Dithorium Pentaselenide, Th_2Se_5   99

　Formation and Preparation   99
　　Synthesis from the Elements   99
　　Enthalpy and Gibbs Free Energy of Formation   99
　Crystallographic Properties   99
　Thermal Properties   100
　Electrical Properties   100
　Chemical Behavior   100

Page

11.1.7   Thorium Triselenide, ThSe$_3$ 100

    Formation and Preparation   100
    Crystallographic Properties   101
    Thermal Properties   101

11.2   Compounds of Thorium with Selenium and Oxygen   102

11.2.1   Thorium Oxide Selenide, ThOSe   102

    Preparation   102
    Crystallographic Properties   102
    Thermal Properties   103

11.3   Compounds of Thorium with Selenium and Nitrogen   104

11.3.1   Dithorium Dinitride Selenide, Th$_2$N$_2$Se   104

11.4   Compounds of Thorium with Selenium Oxoacids   104

11.4.1   Selenites   104

11.4.2   Selenates   105

12   Compounds of Thorium and Tellurium   106

12.1   Binary Thorium Tellurides   106

12.1.1   The Th-Te System   106

12.1.2   Thorium Monotelluride, ThTe   106

    Formation and Preparation   106
    Crystallographic Properties   107
    Thermal Properties   108
    Electrical Properties   108
    Chemical Behavior   108

12.1.3   Dithorium Tritelluride, Th$_2$Te$_3$ 108

    Formation and Preparation   108
    Crystallographic Properties   110
    Thermal Properties   110
    Electrical Properties   110
    Chemical Behavior   110

12.1.4   Thorium Ditelluride, ThTe$_2$ 110

    Formation and Preparation   110
    Crystallographic Properties   111
    Thermal Properties   111
    Electrical Properties   111
    Chemical Behavior   111

Page

12.1.5   Thorium Tritelluride, ThTe₃ 113

Formation and Preparation 113
Crystallographic Properties   113
Thermal Properties 113
Electrical Properties 113
Chemical Behavior 114

12.2   Compounds of Thorium with Tellurium and Oxygen 114

12.2.1   Thorium Oxide Telluride, ThOTe 114

Preparation 114
Crystallographic Properties   115
Thermal Properties 115

12.3   Compounds of Thorium with Tellurium and Nitrogen 115

12.3.1   Dithorium Dinitride Telluride, Th₂N₂Te 115

Preparation 115
Crystallographic Properties   116

12.4   Compounds of Thorium with Tellurium Oxoacids 116

12.4.1   Thorium Tellurites   116

Thorium Tellurite, Th(TeO₃)₂ 116
Thorium Tellurito Complexes, M₅ᴵᴵTh(TeO₃)₇ 117

12.4.2   Thorium Tellurates   117

13   Compounds of Thorium and Boron 118

13.1   Binary Thorium Borides 118

13.1.1   The Th-B System 118

13.1.2   Gaseous Thorium Monoboride 119

13.1.3   Thorium Tetraboride, ThB₄ 119

Formation and Preparation 119
   Synthesis from the Elements 119
   Reduction of ThO₂ with Boron 119
   Special Preparation Techniques 120
   Sintering Behavior and Hot-Pressing 120
   Single Crystals 121
   Enthalpy and Free Energy of Formation 121

Page

Crystallographic Properties and Bonding 121
Mechanical Properties 122
Thermal Properties 122
Electrical Properties 124
    Electric Resistivity 124
    Thermoelectric Power 126
    Hall Coefficient 126
Magnetic Properties 126
Chemical Reactions 127
    On Heating 127
    With Elements 127
    With Compounds 127

13.1.4   Thorium Hexaboride, ThB_6 127

Formation and Preparation 127
    Synthesis from the Elements 128
    Reduction of ThO_2 with Boron 128
    Special Preparation Techniques 128
    Sintering Behavior and Hot-Pressing 129
    Single Crystals 129
    Enthalpy and Free Energy of Formation 129
Crystallographic Properties and Bonding 130
Mechanical Properties 131
Thermal Properties 131
Electrical Properties 133
Magnetic Properties 133
Chemical Reactions 134

13.1.5   Thorium Dodecaboride 134

13.1.6   Thorium Octadecaboride 135

13.1.7   Thorium "Hectoboride" 135

13.2   Ternary Compounds of Thorium and Boron 138

13.2.1   With Hydrogen 138

Thorium Tetrahydroborate, $Th(BH_4)_4$ 138
    Preparation 138
    Crystallographic and Optical Properties 138
    Thermal Properties 139
    Chemical Reactions 139
$LiTh(BH_4)_5$ 140
$Li_2Th(BH_4)_6$ 140
$N(C_4H_9)_4Th(BH_4)_5$ 140

13.2.2   With Sodium 141

Page

13.2.3 With Rare Earth Elements 141

Phase Relationships 141
$Ce_{1-x}Th_xB_4$ Solid Solutions 142
$Ln_{1-x}Th_xB_6$ Solid Solutions 142
    Preparation 142
    Crystallographic Properties 143
    Thermal Stability 144
    Thermionic Emission 144
    X-Ray Absorption Spectrum 144
    Superconductivity 144
    Magnetic Susceptibility 145

13.2.4 With Other Metals 146

$Th_2Fe_{14}B$ 146
Ternary Thorium Compounds with $ThMoB_4$ Structure 147

Table of Conversion Factors 148

10   Compounds of Thorium and Sulfur

Horst Wedemeyer
Kernforschungszentrum Karlsruhe
Institut für Material- und Festkörperforschung
Karlsruhe, Federal Republic of Germany

10.1   Binary Thorium Sulfides

10.1.1   The Th-S System

The binary compounds ThS, Th_2S_3, Th_7S_{12}, ThS_2, and Th_2S_5 are reported to exist in the thorium-sulfur system, based on X-ray diffraction measurements or single crystal determination. A further compound, $ThS_{2.36}$ (or Th_3S_7), prepared by the reaction of the elements [1], has not been confirmed. A compound with composition Th_4S_7, prepared by decomposition of ThS_2 [2], has been shown to be a compound of the formula $Th_{7-x}S_{12}$ by X-ray determination on a single crystal [3].

The solubility of sulfur in thorium was determined to be less than 1 at.% at 1050°C. No intermediate phases were found to exist between Th and ThS, but a eutectic composition was observed with 9.5 at.% sulfur at 1630 ± 15°C [4]. The cubic (NaCl type) thorium monosulfide, silvery in color, has no [5] or a very small homogeneity range [6 to 9]. The melting point of ThS was measured to be 2335°C [10]. A eutectic composition which melts at about 1800°C was found to exist between the phases ThS and Th_2S_3 [6]. A melting point of about 1900°C was found in [11]. The orthorhombic Th_2S_3 (Sb_2S_3 type), brown in color, has no homogeneity range [5]. Its melting point was determined to be 1950 ± 50°C [6]. The hexagonal Th_7S_{12} (Th_7S_{12} type), black in color, has a homogeneity range with S:Th ratios of 1.71 (stoichiometric compound) to 1.76, resulting in the formula $Th_{7-x}S_{12}$ ($0 \leqslant x \leqslant 0.2$) [3]. The melting point of Th_7S_{12} was determined to be 1770 ± 30°C [6]. A eutectic composition which melts at 1765 ± 25°C was found to exist between the phases Th_7S_{12} and ThS_2 [2]. The orthorhombic ThS_2 ($PbCl_2$ type), brown to purple in color, has no homogeneity range [5]. Its melting point was determined to be 1905 ± 30°C [6]. The orthorhombic Th_2S_5 (space group $Pcnb$-D_{2h}^{14}, No. 60), red in color, is reported to dissociate at about 600°C [12].

A tentative phase diagram, based on the measured melting points of the binary compounds and the observed eutectic compositions, is given in **Fig. 1**, p. 2. Within this diagram it is assumed that Th_7S_{12} decomposes peritectically and the eutectic between Th_7S_{12} and ThS_2 is placed at 2100 K [13].

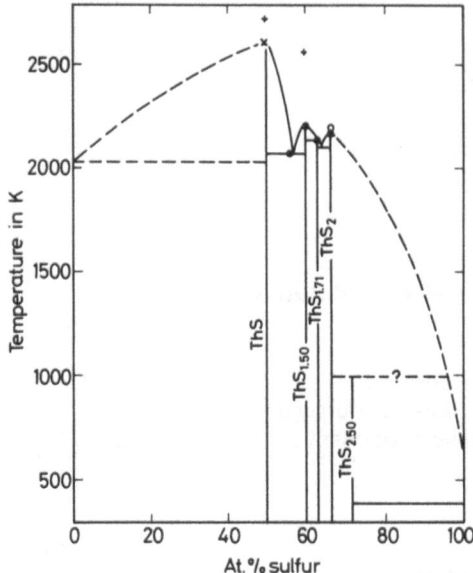

Fig. 1. Tentative phase diagram for the system thorium-sulfur [13], based on: ○ melting temperature of ThS_2 [122]; ● melting temperatures given by [6]; + melting temperatures of ThS and Th_2S_3 [14]; × melting temperature of ThS [22].

10.1.2   Thorium Monosulfide, ThS

10.1.2.1   Formation and Preparation

A face-centered cubic thorium sulfide with composition $ThS_{0.75}$ was prepared in 1941 by the reaction of thorium metal with pure sulfur at 880 to 900°C, placed in a reaction tube and sealed in a quartz tube [1]. This phase composition has not been confirmed as a monosulfide phase as shown by [5, 6]. Newer preparative techniques, commonly used, are based on the reaction of thorium or thorium hydride with hydrogen sulfide or on the reduction of higher sulfides (ThS_2).

Synthesis from the Elements

Stoichiometric amounts of powdered thorium metal and pure sulfur, placed in porcelain boats, were heated under a stream of H_2S [14] according to [15, 16], also cited in [17]. The reaction was carried out at 400 to 850°C with an optimum temperature of 600°C for 30 min. The obtained product, $ThS_{0.98}$, contained considerable amounts of oxygen as an impurity. Highly reactive thorium powder, prepared by hydriding thorium metal at 600°C in a first step, then at 350°C to form the highest thorium hydride, followed by decomposition of the hydride at 700°C in a vacuum of 10^{-5} Torr, was used for this reaction [4, 18]. A stoichiometric amount of thorium powder and pure sulfur was sealed in a quartz tube and heated to 800°C for two weeks [18], see also [17], or to 700°C for one week [4]. The product was thoroughly ground under dry argon, reevacuated, and heated to 800°C for another 7 days, but the monosulfide was found only in a small portion of the sample; mostly ThS_2 and Th_7S_{12} were formed as shown by X-ray diffraction measurement [18]. ThS, with a small amount of ThOS (see p. 48)

[19], was obtained after homogenizing the product at 1600°C in a tungsten crucible in vacuum [4], or after heating the reaction product inductively to between 1400 and 2100°C for 10 min to 9 h using a tantalum tube [19], see also [20].

Reaction of Thorium or Thorium Hydride with H_2S

The reaction of finely divided thorium powder (prepared by hydriding thorium turnings, etc., and thermally decomposing the hydride) [10, 22, 26], also reported in [23, 24], or thorium hydrides [6, 25], also reported in [2, 17, 23], with stoichiometric amounts of H_2S leads to the formation of a mixture of higher thorium sulfides, thorium metal, and some amounts of thorium hydrides. The reaction was carried out in a quartz apparatus (which is shown in "Uranium" Suppl. Vol. C10, 1984, p. 3, Fig. 3) at 700°C for the reaction with thorium powder [10, 22, 26], at 350°C within 20 min using Th_4H_{15} [25], or at 400 to 500°C within about 1 h using $ThH_{3.5}$ [6]. The reaction of thorium powder with H_2S is reported to be somewhat complicated by the existence of two stable thorium hydrides: ThH_2 and Th_4H_{15} [10, 26], see also [27] (compare "Thorium" Erg.-Bd. C1, 1978, p. 2). This reaction product was homogenized at higher temperatures to obtain a uniform product. The reaction mixture was ground to a fine powder (to pass a −100 mesh [22] or a 0.044 mm [25] sieve) in inert dry boxes to prevent contamination with oxygen. The ground powder was heated up first to 500 to 600°C to decompose the contained thorium hydride [6] and then to 1800 to 1900°C in an inert atmosphere or in a vacuum ($<2 \times 10^{-4}$ Torr [22]) using molybdenum crucibles [6, 10, 22, 26] or tungsten crucibles [24], see also [17, 27, 28]. The homogenized product contained about 97 wt% ThS [10, 22, 26] with ThOS as the main impurity. A not quite homogeneous ThS was obtained after homogenization in the molten state at about 2000°C in vacuum [29]. Very pure ThS was obtained after homogenization at 1850°C in a vacuum of 10^{-6} Torr within 4 h, regrinding (in a boron carbide mortar) and homogenization at 1850°C for 8 h in vacuum (impurities in ppm: C = 88, H = 4, N = 15, O = 950; S/Th = 1.043 ± 0.005) [25].

Reduction of ThS_2

To prepare ThS, mixtures of ThS_2 and thorium hydride in proper amounts were heated first to 400 to 600°C under reduced pressure to decompose the hydride to finely divided thorium metal and then heated to 2000 to 2200°C in a vacuum using a molybdenum container [6], see also [17]. The reaction of ThS_2 with ThH_4 carried out in a vacuum started at about 200°C and was completed at 1500°C [30]. According to [6], a stoichiometric mixture of ThS_2 and thorium powder, pressed into pellets, was heated up to 1650°C for 1 h to form ThS [31], see also [9]. Electron beam melting of ThS_2 and thorium metal in proper amount resulted in ThS with an S/Th ratio of 1.0 [8].

Reduction of ThOS

ThS was obtained by the reduction of ThOS with aluminium according to 3 ThOS + 2 Al = 3 ThS + Al_2O_3. The reaction was carried out at 1150°C and the Al_2O_3 formed was sublimed at 1750°C in vacuum [32].

Reaction of Thorium with ZnS

ThS was prepared by the reaction of thorium metal powder with zinc sulfide, ZnS, in stoichiometric amounts. The powder, thoroughly mixed by grinding, pressed into pellets and placed in a tantalum boat, was heated to 600°C in a silica tube under a stream of pure argon. After the reaction was completed, the reaction tube was evacuated and heated to 1500°C to remove any traces of the volatile zinc (boiling point at 906°C). The ThS obtained by this procedure was a grayish powder stable in air (Th = 88.13%, S = 11.07%, O_2 = 0.71%, N_2 = 0.01%). The level of oxygen-containing impurities depended mostly on the purity of the zinc sulfide [33].

Fused Salt Electrolysis

ThS was prepared by electrolytic reduction of $ThCl_4$ in an oxygen-free fused salt medium. The reduction was carried out in an electrolytic cell which was also used for the preparation of US (see "Uranium" Suppl. Vol. C 10, 1984, p. 5, Fig. 4). In this cell, a graphite crucible was used as an anode and a rotating molybdenum shaft and paddle as a cathode. The experiment was carried out with 37.4 g of anhydrous $ThCl_4$ and 7.8 g of anhydrous Na_2S mixed and placed in the diaphragm cup. The cell was then filled with a NaCl/KCl salt mixture in eutectic composition (melting point 658°C). The electrolytic reduction was carried out at 800°C passing a current of 4 to 6 A at 4.4 to 5.0 V through the cell for 150 min (11.0 Ah), which is twice the theoretical requirement for the reduction to dipositive thorium ions. The product was then vacuum-distilled to remove the salt. A yield of 78% was obtained by this procedure [34, 35], see also [17, 23].

Sintering Behavior

ThS powder can be sintered in vacuum or in argon atmosphere to densities of 97 to 98 of theoretical density (th. d.) [22, 26].

Pressed powders containing 97 to 98 wt% of the monosulfide phase showed a rapid increase of the density from 86 to 98% th. d. at increasing temperatures from 1900 to 2000°C, when sintered in vacuum. The oxygen impurities, present as ThOS, are stable at temperatures up to 1950°C [22, 26], when a ThS-ThOS eutectic is formed, which leads to liquid sintering as shown by polished sections of the sintered bodies [22]. The oxide sulfide phase was found at the grain boundaries and also as needle-like precipitates on certain crystal planes, probably the (111) plane [10, 22, 26], see also [27, 36, 37], also cited in [17, 38, 39]. It should be noted that ThOS was suggested to be unstable in the presence of ThS, reacting to give ThO_2 and Th_2S_3 as observed from ThO_2-saturated samples of ThS ($ThS_{0.98}$ · 0.06 ThO_2) [6]. Sintered bodies, preformed at 15000 psi and isostatically hot-pressed at 55000 to 85000 psi, showed metallic reflectance and were silvery in color [40], see also [41]. ThS powder compacts supported on Mo sintered at 10^{-3} to 10^{-5} Torr gave highly refractory crucibles. The powders were ground in inert atmospheres to pass a 200 or 325 mesh screen, and pressed at 50000 to 100000 psi with naphthalene added as a binder. The rate of recrystallization was observed to be quite slow below 2000°C [42], see also [43, 44]. Further results of hot-pressed ThS powders at a pressure of 350 kg/cm^2 are summarized in Table 1.

Table 1
Sintering of Hot-Pressed ThS Powders at a Pressure of 350 kg/cm^2 [14].

sintering temperature in °C	time of sintering in min	density in g/cm^3	residual porosity in %
1810	5	5.82	34.8
2030	8	8.53	4.2
2160	8	7.47	16.1
2230	melted	—	—

Purification of ThS

ThS containing ThO_2 and ThOS in minor amounts was purified by heating up to a temperature of about 2200 °C at reduced pressure ($< 10^{-3}$ Torr) in the absence of any reactive material to form a molten bath and to vaporize the ThO_2 and ThOS without vaporizing substantial amounts of ThS [45].

Single Crystals

Single crystals of ThS were prepared by electron beam melting of pure ThS powders, and subsequently annealing at 1900 °C for 15 h in vacuum (1×10^{-5} Torr). Single crystals of about 1 cm^3 in volume were obtained by this procedure. The cleavage plane was the (100) plane. Measured angles between the cleavage plane and several other crystallographic planes are: (511) 16.1°, (311) 25.4°, (310) 18.8°, (210) 26.7° [8].

Enthalpy, Entropy, and Gibbs Free Energy of Formation

The enthalpy value ΔH_f (298 K) = -120 ± 5 kcal/S-atom was estimated by [2], also cited in [6, 13, 14, 17, 46], see also [47, 48]. This value was evaluated to ΔH_f° (298 K) = -105 ± 10 kcal/mol in [52]. The value ΔH_f (298 K) = -100 ± 10 kcal/mol, estimated by [116] has been also cited in [52]. Measurements with solid electrochemical cells of the type Th, $ThF_4/CaF_2/ThF_4$, ThS, Th_α in the range 973 to 1173 K gave ΔH_f° (1173 K) = -102.2 kcal/g-atom S [9], when the enthalpy of formation of Th_2S_3 [50] was used as an auxiliary value, or ΔH_f° (1173 K) = -109.0 kcal/g-atom S [9], when the entropy of vaporization of ThS from [51] was used, see also [13]. ΔH_f° (c, 298 K) = -93.4 kcal/mol is the value selected by [52], introducing the reaction Th(c) + Th_2S_3(c) = 3 ThS(c), see also [13]. The enthalpy of formation was found to be ΔH_f° ($ThS_{1.03}$, c, 298 K) = -96.6 kcal/mol based on HCl solution calorimetry according to ThS(c) + 4 HCl (aq) = $ThCl_4$(aq) + H_2S(aq) + H_2(g) [53], also cited in [13, 54], see also [55, 56], or $-95.5 + 1.5$ kcal/mol [57], see also [13], or -94.54 kcal/mol [58], see also [13].

The entropy of formation, ΔS_f°, was calculated using the entropy of thorium metal (53.39 $\pm$ 0.27 J · mol^{-1} · K^{-1}) and of rhombic sulfur (31.88 $\pm$ 0.16 J · mol^{-1} · K^{-1}) to be ΔS_f° (298.15 K) = -15.46 J · mol^{-1} · K^{-1} [25]. A tentative value for the Gibbs free energy of formation, derived from this value together with ΔH_f° = -435 kJ/mol (from [9]), is given to be ΔG_f° = -430 kJ/mol [25]. A value of ΔG_f°(298 K) = -96.0 kcal/mol was calculated from ΔH_f° (at 298 K) = -97.1 kcal/mol [55] and ΔS_f° = -3.69 cal · mol^{-1} · K^{-1} (from [25]). Further estimated values of ΔS_f° and ΔG_f° are given in [59].

10.1.2.2 Crystallographic Properties, Bonding, and Lattice Dynamics

Thorium monosulfide crystallizes in the face-centered cubic system (NaCl type) with 4 molecules per unit cell; the space group is $Fm3m\text{-}O_h^5$ (No. 225) [60]. The lattice constant of ThS depends on the sulfur content, varying between 5.679 Å for $ThS_{0.92}$ and 5.683 Å for $ThS_{1.0}$ (see **Fig. 2**, p. 6) [8]. Lattice parameters of ThS in the presence of thorium metal or Th_2S_3 are given by [9] and summarized in Table 2, p. 6.

                     References for 10.1 on pp. 45/8

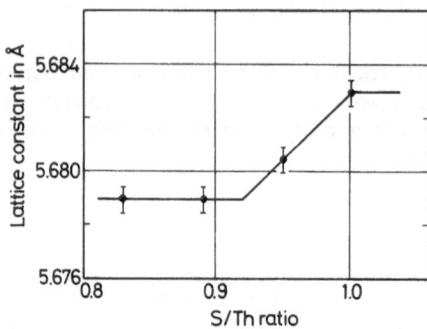

Fig. 2. Change of the lattice parameter a with the composition ratio S/Th [8].

Table 2

Measured Lattice Parameters of ThS in Presence of Thorium Metal or Th_2S_3 [9].

nominal composition, S/Th ratio	lattice parameter a in Å ± 0.002 Å	oxygen content in wt% ± 0.10 wt%
0.70	5.678	
0.75	5.680	0.67
0.80	5.678	0.54
1.00	5.687	0.51
1.00	5.681	0.64
1.20	5.674	0.49
1.25	5.675	0.56
1.30	5.674	

Different measured values of the lattice constant, which are reported in the literature, are summarized in Table 3.

The X-ray density, calculated from X-ray diffraction measurements, is given as 9.56 [5], 9.54 [26], 9.51 [38] g/cm³. The interatomic bond distance Th-S is 2.84 Å [6] or 2.841 Å [61]. The ionic radii, calculated (cation radius) and observed from cation-anion separation experiments are: anion radius (S^{2-}) = 1.84 Å [62] or 1.844 Å [61], cation radius (Th^{4+}) = 0.99 Å [62] or 0.997 Å [61]. It was determined from X-ray diffraction measurements (Laue pattern) that the cleavage surface of ThS single crystals is the (100) plane [8].

High pressure (up to ca. 40 GPa) X-ray diffraction studies performed on ThS using synchrotron radiation indicate a phase transformation to ThS II above 15 to 20 GPa. The resulting structure could be described as distorted face-centered cubic with the hexagonal lattice parameters of a = 380 ± 3 pm (a = $a_{fcc}/\sqrt{2}$) and c = 1861 ± 20 pm (c = $2a_{fcc} \cdot \sqrt{3}$) (see **Fig. 3** and **Fig. 4**, p. 8) [63]. The isothermal bulk modulus, B_0, and its pressure derivative at ambient pressure, B_0', were calculated from Murnaghan's equation p = (B_0/B_0') $[(V_0/V)^{B_0'} - 1]$ and the Birch equation p = $^3/_2 B_0 [(a_0/a)^7 - (a_0/a)^5][1 + ^3/_4 (B_0' - 4)\{(a_0/a)^2 - 1\}]$. The results are summarized in Table 4, p. 8. Measurements up to 40 GPa gave the same results [140].

Table 3
Measured Lattice Parameters of ThS.
References cited in the "Ref." column after the semicolon are secondary ones. Values in Å which are recalculated from the original values in kX are given in parentheses.

lattice constant a	method of preparation	Ref.
5.671 ± 0.002 kX (= 5.682 ± 0.0021 Å)	$ThS_2 + ThH_x$, $ThH_x + H_2S$	[5]; [6, 12, 13, 17, 18, 64, 65, 66]
5.669 to 5.671 kX (= 5.673 to 5.681 Å)	in presence of Th	[6]; [13, 67]
5.685 Å	Th (ex ThH_x) + H_2S	[21]; [13, 17, 23, 27, 36, 65, 68]
5.686 Å	Th + H_2S	[38]; [69]
5.682 Å	Th (ex ThH_x) + H_2S; ThS with 5% ThOS	[70]
5.6809 ± 0.0009 Å	Th (ex ThH_x) + H_2S; $ThS_{1.01}$	[51]; [13, 67]
5.680 Å		[49]; [31, 71]
5.671 Å	Th + ZnS; 88.13% Th, 11.07% S, 0.71% O, 0.01% N	[33]
5.679 to 5.681 Å	Th (ex ThH_x) + ThS_2; $ThS_{0.92}$ to $ThS_{1.00}$	[7]
5.683 Å	single crystal	[7]
5.671 Å		[60]
5.68 Å	$ThS_{0.98}$; 2% ThOS	[72]
5.681 ± 0.002 Å	Th + S	[20]; [19]
5.682 ± 0.002 Å	Th + S; some ThO_2	[19]
5.683 ± 0.001 Å	Th + ThS_2	[73]; [13]
568.51 ± 0.03 pm	Th + S	[63]

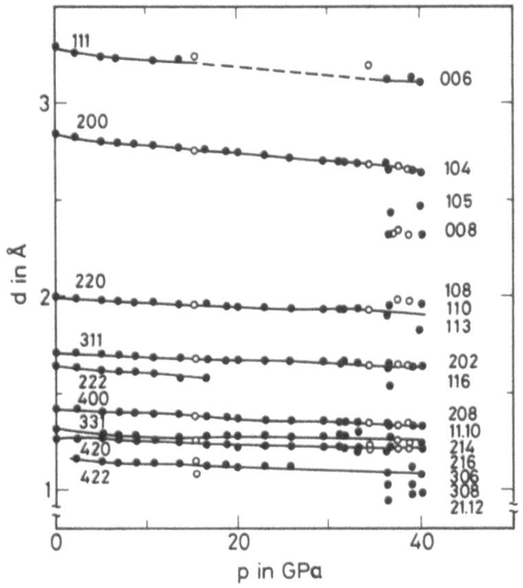

Fig. 3. Interplanar spacings of ThS as functions of pressure [63]. ● increasing pressure; ○ decreasing pressure. The indices of the fcc phase are given on the left-hand side, those of the distorted fcc phase are given on the right-hand side.

References for 10.1 on pp. 45/8

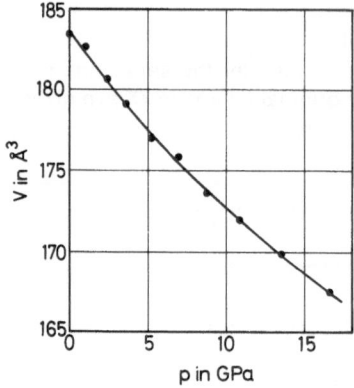

Fig. 4. Unit cell volume of ThS as a function of pressure [63].

Table 4
Bulk Modulus and its Pressure Derivative for ThS [63].

Murnaghan equation	$B_0 = 145.7 \pm 5.3$ GPa
	$B_0' = 5.2 \pm 1.0$
Birch equation	$B_0 = 145.2 \pm 5.6$ GPa
	$B_0' = 5.4 \pm 1.0$
average value	$B_0 = 145 \pm 6$ GPa
	$B_0' = 5.4 \pm 1$
from [74]	$B_0 = 220$ GPa

The values of B_0 and B_0', determined from the two above equations, are in good agreement. However, these values differ by about 50% from a value of $(4.51 \pm 0.7) \times 10^{-4}$ kbar^{-1} given by [74] for the isothermal compressibility [63].

The interatomic potentials for a large number of metal chalcogenide crystals, including ThS, were calculated using a logarithmic interaction potential energy function. For ThS, a cohesive energy of W = 2960 kJ/mol was obtained by a procedure which made use of the Moelwyn-Hughes parameter C_1. The following related values were calculated: compressibility $\beta_0 = 14.01 \times 10^{12}$ Pa^{-1}, force constant f $= 12.2 \times 10^{-4}$ N/m, IR absorption frequency $\nu_0 = 8.1 \times 10^{12}$ Hz, Debye temperature $\theta_D = 390$ K, Grüneisen parameter $\gamma = 1.73$, Anderson-Grüneisen parameter $\delta = 3.46$, Moelwyn-Hughes parameter $C_1 = 4.46$ [75]. A value of A $= 6.33 \times 10^5$ dyn/cm was determined for the nearest-neighbor force constant in ThS from neutron scattering experiments, using time-of-flight spectrometry [72].

Calculations within the framework of the Born model gave a cohesive energy of W $=$ 844 kcal/mol, when an exponential form was used, and W = 824 kcal/mol, when an inverse power form was used. The zero point energy was 1.74 kcal/mol. The following repulsive energies were calculated: 168 kcal/mol (exponential form) and 181 kcal/mol (inverse power form) for cation-anion interaction, 0.23 kcal/mol (exponential form) and 0.90 kcal/mol (inverse power form) for cation-cation interaction, 7.9 kcal/mol (exponential form) and 14.8 kcal/mol (inverse power form) for anion-anion interaction [76].

10.1.2.3 Mechanical Properties

Density

For X-ray densities of ThS, see p. 6. Experimental densities of 86 to 98% th.d. were obtained from sintered specimens (see Section 10.1.2.1 "Formation and Preparation" p. 2).

Elasticity, Hardness, and Strength

Third-order elastic constants were calculated using the Born-Mayer potential model. The results for ThS at 0 K were (in 10^{12} dyn/cm^2): $C^o_{111} = -23.039$, $C^o_{112} = -2.480$, $C^o_{123} = 0.962$ [77].

The Vickers hardness was measured for different ThS samples (85 to 98% th.d.) at a load of 100 g. The following results are reported: 234 kg/mm^2 [41], also cited in [22], 165 to 270 kg/mm^2 [26], 219 to 246 kg/mm^2 [36], also cited in [17, 27]. Measured values for the microhardness of different samples are reported: for powder particles, 143.4 ± 29 kg/mm^2 (5 g load); for sintered specimens, 800 ± 81.6 kg/mm^2 (30 g load); and for sintered specimens after annealing, 363 ± 39.8 kg/mm^2 (20 g load) [14].

The modulus of rupture for ThS samples of varying density (85 to 98% th.d.) was measured to be in the range of 16000 to 31000 psi [24], also cited in [17, 22, 27, 36].

10.1.2.4 Thermal Properties

Thermal Expansion

The thermal expansion coefficient, α, of ThS increases slightly with temperature. The results of different measurements are summarized in Table 5, see also **Fig. 5.**, p. 10.

Table 5
Thermal Expansion Coefficient, α, of ThS.

temperature range in °C	α in 10^{-6} K^{-1}	Ref.
0 to 100	9.2	[24]
0 to 500	9.5	[24]
0 to 975	10.2	[24]
up to 1000	10.1	[26], see also [22, 23]
0 to 1000	10.2	[36], see also [17, 27, 65]
measured by high-temperature X-ray diffraction:		
850 to 1400	12.0	[31], see also [49, 78]

Vaporization

Values of the vapor pressure [2, 79] or decomposition pressure [6] of ThS are reported to be $<10^{-3}$ Torr at 2200°C [6], also cited in [2, 14], and 10^{-3} Torr at 1870°C [79]. ThS is as non-volatile as the most refractory oxides, see [79]. An estimation of the total effective pressure, p_e, of ThS (calculated as if the vapor species were all ThS) in the range of 1935 to 2464 K is

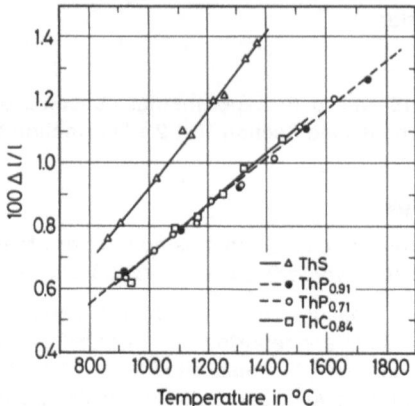

Fig. 5. Thermal expansion of ThS [31]. Phosphides and carbide are shown for comparison.

given by the equation log p_e (in atm) $\approx 7.7 - (3.366 \times 10^4)/T$, based on effusion measurement, collecting the effusate on high purity disks of fused silica and assayed by neutron activation analysis [51], also cited in [23, 66]. The thorium decomposition pressure, p_{Th}, was calculated from the enthalpies of formation of ThS(c), Th(g), and S(g) to be log p_{Th} (in atm) $= 7.43 - 33050/T$ (at 1700 to 3000 K) [57], see also [13]. Assuming the partial pressure of ThS(g) to be one third that of Th(g), the total effective pressure was estimated to be log p_e (in atm) $\approx 7.5 - 33000/T$ (at 1700 to 3000 K) [57], see also [13].

Melting Point

The highest melting point of ThS was measured to be in the range of 2400 to 2450°C (direct observation) [14], see also [17], the lowest one 2200°C (method not reported) [47], see also [80]. Further measured melting points are reported: >2200°C (modified Mendenhall wedge, in hydrogen) [6, 27], see also [2, 24, 40, 59, 65, 81, 82], >2300°C (method not reported) [11], 2330°C (method not reported) [38], see also [69], 2698 K (direct observation, average value) [83], see also [14]. The melting points were measured by different methods; most of the samples contained several % of oxygen in the form of ThOS (see p. 48). In hydrogen, the value was determined to be 2335°C [10], see also [6, 23, 31, 39] or 2340°C (Mendenhall wedge) [37].

Heat Capacity and Thermodynamic Functions

The heat capacity of ThS was obtained in the low-temperature range of 1.5 to 10 K by neutron inelastic scattering [73] and in the range of 1 to 22 K by an isoperibol method [25]. Below 4.2 K, the heat capacity follows the equation C_p (in J · mol^{-1} · K^{-1}) $= 3.70 \times 10^{-3}$ T + 3.98×10^{-4} T^3 (see **Fig. 6**) [25]. The heat capacities, listed in Table 6, are obtained from a sample of the composition ThS$_{1.043 \pm 0.005}$. The measured values were corrected by -0.38%, since the excess sulfur was assumed to be in the form of Th$_2$S$_3$ in the second phase.

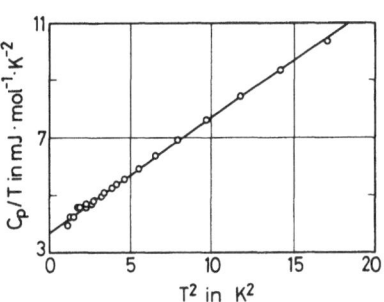

Fig. 6. C_p/T versus T^2 for ThS [25]. Observed points for $ThS_{1.043}$ less a 0.38% correction to stoichiometric ThS. The straight line represents the equation $C_p/T = T^2 + \gamma$.

Table 6
Heat Capacity, C_p, of $ThS_{1.043}$ from 1 to 22 K Measured by an Isoperibol Method. Molecular Weight 265.48 [25].

temperature in K	heat capacity, C_p, in $J \cdot mol^{-1} \cdot K^{-1}$	temperature in K	heat capacity, C_p, in $J \cdot mol^{-1} \cdot K^{-1}$
1.082	0.00433	3.776	0.03541
1.174	0.00502	4.133	0.04304
1.244	0.00532	4.557	0.05367
1.346	0.00621	4.694	0.05770
1.385	0.00639	5.205	0.07420
1.518	0.00700	5.717	0.09329
1.532	0.00728	6.292	0.1197
1.641	0.00778	6.924	0.1554
1.671	0.00804	7.582	0.2026
1.680	0.00811	8.296	0.2659
1.804	0.00896	9.037	0.3505
1.854	0.00941	9.877	0.4683
1.856	0.00949	10.793	0.6289
1.980	0.01046	11.458	0.7653
2.041	0.01104	11.813	0.8460
2.052	0.01112	12.454	1.007
2.162	0.01209	13.582	1.334
2.353	0.01399	14.912	1.782
2.567	0.01643	16.443	2.382
2.828	0.01965	18.205	3.157
2.121	0.02385	20.045	4.047
2.441	0.02909	21.887	5.008

In the general equation $C_p/T = \alpha T^2 + \gamma$, the coefficient α and the term γ (apparent electronic specific heat), derived from neutron inelastic scattering, are $\alpha = 383\ \mu J \cdot mol^{-1} \cdot K^{-4}$ and $\gamma = 3.9\ mJ \cdot mol^{-1} \cdot K^{-2}$ [73], see also [7, 84]. The heat capacity of ThS was measured with an adiabatic calorimeter in the range 8 to 350 K [25] and 5 to 300 K [85]. Measured values

References for 10.1 on pp. 45/8

Table 7
Heat Capacity, C_p, of $ThS_{1.043}$ from 8 to 350 K Measured by an Adiabatic Method. Below 20 K, C_p values in this table are less accurate than those in Table 6 [25].

temperature in K	heat capacity in $cal \cdot mol^{-1} \cdot K^{-1}$	temperature in K	heat capacity in $cal \cdot mol^{-1} \cdot K^{-1}$
8.03	0.248	100.98	29.95
9.74	0.451	110.41	31.95
11.59	0.790	120.25	33.88
13.73	1.384	130.08	35.60
15.76	2.102	140.11	37.18
17.73	2.952	150.30	38.61
19.77	3.916	160.49	39.87
21.89	4.984	170.56	41.00
24.16	6.179	180.56	42.05
27.35	7.875	190.55	42.93
29.64	9.055	200.55	43.71
32.11	10.26	210.54	44.37
34.81	11.49	220.54	45.05
37.93	12.82	230.33	45.72
41.51	14.21	240.41	46.27
41.95	14.33	250.50	46.71
46.15	15.83	260.59	47.10
50.85	17.31	270.64	47.62
56.04	18.86	280.66	48.09
61.67	20.40	290.65	48.48
67.76	22.03	300.60	48.80
74.35	23.75	310.48	49.04
81.51	25.55	320.30	49.35
85.25	26.43	330.12	49.59
89.33	27.46	340.03	49.84
92.60	28.21	347.54	50.05

corrected to pure ThS are summarized in Table 7, which are in good agreement with the values given in [85]. The heat capacity, calculated from measured characteristic temperatures, $\theta_{Th} = 168.8$ K and $\theta_S = 498.9$ K, is given by the equation C_p (in $cal \cdot mol^{-1} \cdot K^{-1}$) $= 11.80 + 1.20 \times 10^{-3}\, T - 0.64 \times 10^5/T^2$ [83], see also [13].

Further estimates are presented for the temperature range of 298 to 2000 K by [52], see also [13]: C_p (in $cal \cdot mol^{-1} \cdot K^{-1}$) $= 11.10 + 1.4 \times 10^{-3}\, T$ or by [57], see also [13] (see Table 8): C_p (in $cal \cdot mol^{-1} \cdot K^{-1}$) $= 11.98 + 1.305 \times 10^{-3}\, T - 0.857 \times 10^5/T^2$. The results of the different estimates are shown in **Fig. 7**.

Values for the entropies, enthalpies, and Gibbs free energies, derived from the measured heat capacities, are summarized in Table 9, p. 15. Data at 298 K are: $C_p^{\circ} = 47.72 \pm 0.24$ $J \cdot mol^{-1} \cdot K^{-1}$, $S^{\circ} = 69.81 \pm 0.35\ J \cdot mol^{-1} \cdot K^{-1}$, $H^{\circ} - H_0^{\circ} = 9700 \pm 49$ kJ/mol, $-(G^{\circ} - H_0^{\circ})/T = 37.21 \pm 0.19\ J \cdot mol^{-1} \cdot K^{-1}$ at 298.15 K [25], $C_p = 48.3 \pm 0.5\ J \cdot mol^{-1} \cdot K^{-1}$, $S = 70.6 \pm 0.7$ $J \cdot mol^{-1} \cdot K^{-1}$, $H_T - H_0 = 9840 \pm 100$ kJ/mol, $H_0 - G_T = 11.23 \pm 12$ kJ/mol at 298 K [85], see also [13].

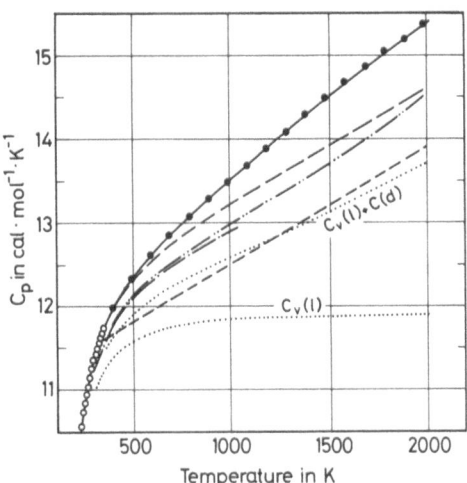

Fig. 7. Heat capacity of ThS(c) [13]. ○ low-temperature results of [25] corrected to stoichiometric ThS; — · — · — estimated by [83]; · · · · · estimated by [52]; ----- estimated by [57]; —●—●— calculated according to $C_v(l) + C(d) + C(e)$ [13]; — · · — calculated according to $C_v(l) + C(d) + \frac{1}{2} C(e)$ [13].
l = lattice, d = dilation, e = electronic.

Values of the entropy of ThS were calculated from characteristic temperatures, measured by X-ray diffraction methods, using a heavy mass model (θ_{Th} = 176.8 K) or an average mass model (θ = 233 K) to be S° (heavy mass model) = 16.9 cal · mol^{-1} · K^{-1} or S° (average mass model) = 19.3 cal · mol^{-1} · K^{-1} [71], see also [46, 91]. These values are comparable with S° = 18.0 cal · mol^{-1} · K^{-1} estimated from a comparison of actinide compounds [86], also cited in [46]. Further estimated values of the entropy are given in [59].

A value of the dissociation energy D°(ThS, 0) = 136 ± 10 kcal/mol follows from the assumption that the partial pressure of ThS(g) is one third that of Th(g) [13, 57]. From this, the standard enthalpy of formation is ΔH_f°(ThS,g,298) = 72 ± 10 kcal/mol [13]. Detailed data of the thermodynamic functions of gaseous ThS are given in [13], see Table 10, p. 16.

Table 8
Thermodynamic Functions of Stoichiometric Thorium Monosulfide, ThS [25].

T in K	C_p° in J·mol^{-1}·K^{-1}	S° in J·mol^{-1}·K^{-1}	H°−H$_0^\circ$ in J/mol	−(G°−H$_0^\circ$)/T in J·mol^{-1}·K^{-1}
5	0.0666	0.0349	0.1075	0.0134
10	0.4852	0.1735	1.2200	0.0515
15	1.810	0.5853	6.5186	0.1507
20	4.011	1.3921	20.805	0.3519
25	6.610	2.5629	47.280	0.6717
30	9.200	4.0000	86.890	1.1037

Table 8 (continued)

T in K	C_p° in $J \cdot mol^{-1} \cdot K^{-1}$	S° in $J \cdot mol^{-1} \cdot K^{-1}$	$H^\circ - H_0^\circ$ in J/mol	$-(G^\circ - H_0^\circ)/T$ in $J \cdot mol^{-1} \cdot K^{-1}$
35	11.54	5.5961	138.81	1.6301
40	13.56	7.2717	201.66	2.2301
45	15.35	8.9743	274.03	2.8848
50	16.95	10.676	354.85	3.5789
60	19.79	14.023	538.88	5.042
70	22.36	17.269	749.78	6.558
80	24.78	20.414	985.56	8.095
90	27.10	23.468	1245.1	9.634
100	29.29	26.437	1527.1	11.166
110	31.33	29.326	1830.3	12.687
120	33.19	32.133	2153.1	14.191
130	34.88	34.857	2493.6	15.676
140	36.40	37.499	2850.1	17.141
150	37.78	40.058	3221.1	18.584
160	39.01	42.537	3605.2	20.004
170	40.13	44.936	4001.0	21.401
180	41.13	47.258	4407.4	22.773
190	42.02	49.506	4823.2	24.121
200	42.81	51.682	5247.4	25.445
210	43.51	53.788	5679.0	26.745
220	44.13	55.827	6117.3	28.021
230	44.71	57.801	6561.5	29.273
240	45.24	59.715	7011.3	30.502
250	45.74	61.572	7466.2	31.707
260	46.22	63.376	7926.1	32.891
270	46.67	65.129	8390.6	34.053
280	47.09	66.834	8859.4	35.193
290	47.46	68.493	9332.2	36.313
300	47.78	70.107	9808.4	37.413
310	48.07	71.679	10288	38.493
320	48.35	73.209	10770	39.554
330	48.61	74.701	11255	40.596
340	48.85	76.156	11742	41.621
350	49.08	77.575	12232	42.628
273.15	46.81	65.67	8538	34.41
298.15	47.72	69.81	9720	37.21
	±0.24	±0.35	±49	±0.19

Thermal Conductivity

The thermal conductivity of ThS was measured by a comparative method [87] based on the Powell comparator [88], see **Fig. 8**, p. 17. Further single values at 75 and 300°C are summarized in Table 11, p. 17. At 25°C, a value of 0.11 cal $\cdot$ cm^{-1} $\cdot$ s^{-1} $\cdot$ K^{-1} was measured [23].

Table 9

Thermodynamic Functions of Thorium Monosulfide, ThS(c), at Higher Temperatures [13].

In the original paper [13] the units for this table inadvertently have been given with J, but should be expressed with cal. A table of the same thermodynamic functions in S I units (based on Joule) is also given in [13].

T in K	C_p° in cal·mol^{-1}·K^{-1}	S° in cal·mol^{-1}·K^{-1}	$-(G_T^\circ - H_{298}^\circ)/T$ in cal·mol^{-1}·K^{-1}	$H_T^\circ - H_{298}^\circ$ in cal/mol	ΔH_f° in cal/mol	ΔG_f° in cal/mol	log K_p
298	11.410	16.680	16.680	0	− 95500	− 94385	69.185
300	11.420	16.751	16.680	21	− 95501	− 94378	68.753
400	12.000	20.121	17.136	1194	− 96091	− 93973	51.344
500	12.330	22.836	18.013	2411	− 96501	− 93397	40.823
600	12.570	25.106	19.011	3657	− 96819	− 92741	33.781
700	12.760	27.059	20.025	4924	− 97079	− 92044	28.737
800	12.920	28.773	21.013	6208	−110301	− 92508	25.272
900	13.070	30.303	21.962	7507	−110208	− 90289	21.925
1000	13.220	31.688	22.866	8822	−110123	− 88081	19.250
1100	13.350	32.954	23.727	10150	−110048	− 85881	17.063
1200	13.490	34.122	24.545	11492	−109982	− 83687	15.241
1300	13.630	35.207	25.324	12848	−109925	− 81498	13.701
1400	13.760	36.222	26.067	14218	−109876	− 79313	12.381
1500	13.900	37.176	26.776	15601	−109837	− 77131	11.238
1600	14.030	38.077	27.454	16997	−109807	− 74952	10.238
1700	14.170	38.932	28.104	18407	−110600	− 72740	9.351
1800	14.310	39.746	28.729	19831	−110527	− 70516	8.562
1900	14.460	40.524	29.329	21270	−110470	− 68294	7.856
2000	14.600	41.269	29.908	22723	−110431	− 66076	7.220

References for 10.1 on pp. 45/8

Table 10
Thermodynamic Functions of Gaseous Thorium Monosulfide, ThS(g) [13]. See note in Table 9 on p. 15.

T in K	C_p° in cal·mol⁻¹·K⁻¹	S° in cal·mol⁻¹·K⁻¹	$-(G_T^\circ - H_{298}^\circ)/T$ in cal·mol⁻¹·K⁻¹	$H_T^\circ - H_{298}^\circ$ in cal/mol	ΔH_f° in cal/mol	ΔG_f° in cal/mol	log K_p
298	8.250	60.310	60.310	0	72000	60107	−44.059
300	8.256	60.361	60.310	15	71993	60033	−43.734
400	8.515	62.774	60.637	855	71070	56126	−30.666
500	8.664	64.691	61.263	1714	70302	52478	−22.938
600	8.788	66.281	61.971	2586	69610	48983	−17.842
700	8.947	67.647	62.686	3472	68969	45593	−14.235
800	9.178	68.855	63.383	4378	55369	41096	−11.227
900	9.490	69.954	64.053	5311	55096	39329	− 9.550
1000	9.870	70.973	64.695	6278	54833	37591	− 8.215
1100	10.288	71.933	65.309	7286	54588	35878	− 7.128
1200	10.709	72.847	65.900	8336	54362	34187	− 6.226
1300	11.103	73.720	66.468	9427	54154	32515	− 5.466
1400	11.444	74.556	67.016	10555	53961	30858	− 4.817
1500	11.719	75.355	67.545	11714	53776	29214	− 4.256
1600	11.922	76.118	68.058	12897	53593	27583	− 3.768
1700	12.054	76.845	68.553	14096	52589	25997	− 3.342
1800	12.121	77.536	69.033	15305	52447	24436	− 2.967
1900	12.132	78.192	69.498	16518	52278	22885	− 2.632
2000	12.098	78.814	69.949	17730	52076	21342	− 2.332
2100	12.030	79.402	70.385	18937	48434	19939	− 2.075
2200	11.935	79.960	70.808	20135	48046	18591	− 1.847
2300	11.822	80.488	71.217	21323	47646	17262	− 1.640
2400	11.697	80.989	71.614	22499	47231	15949	− 1.452
2500	11.566	81.463	71.998	23662	46802	14655	− 1.281
2600	11.433	81.914	72.371	24812	46358	13377	− 1.124
2700	11.300	82.343	72.733	25949	45900	12117	− 0.981
2800	11.170	82.752	73.083	27072	45426	10876	− 0.849
2900	11.045	83.142	73.423	28183	44938	9650	− 0.727
3000	10.925	83.514	73.754	29281	44436	8442	− 0.615

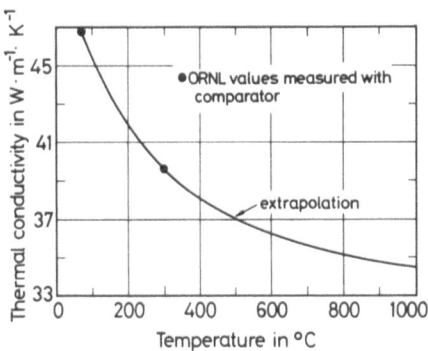

Fig. 8. Thermal conductivity of ThS, corrected to 100% density [87].

Table 11
Thermal Conductivity of ThS.

temperature in °C	thermal conductivity in $W \cdot m^{-1} \cdot K^{-1}$	remarks	Ref.
75	44.8	95.59% th.d.	[89]
75	46.8	corrected to 100% th.d.	[89]
300	34.3	95.59% th.d.	[89]
300	35.9	corrected to 100% th.d.	[89]
300	37.3		[90]

10.1.2.5 Electrical Properties

Electronic Structure

A proposed electronic band structure for ThS is given in **Fig. 9**, which is based on the measured transport properties of ThS. These are: the number of electrons per thorium atom which was calculated from the Hall constant (see p. 20) by a free electron model; the metallic character of the electrical resistivity (see [7, 70], p. 18); the negative thermoelectric power,

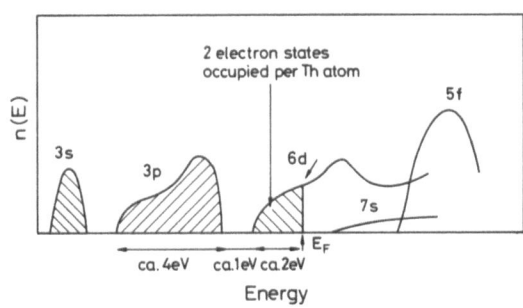

Fig. 9. Schematic band structure of ThS [73].

References for 10.1 on pp. 45/8

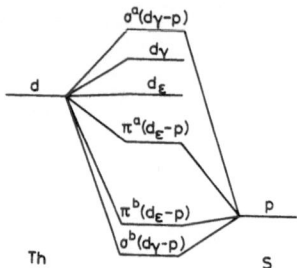

Fig. 10. Schematic diagram of the energy levels for ThS [91]. a = antibonding, b = bonding.

see [70] (p. 19), from which a Fermi energy of 3 eV was calculated and the density of states at the Fermi level estimated to be 0.5 states/eV per Th atom and spin assuming two electron states per Th atom occupied in the band [73], see also [30]. The band structure of ThS was also calculated by a tight-binding method, using a two-center model [91]. From this, the occupied states consist mainly of σ-type bonding between p (sulfur) and d_γ (thorium) electrons as well as π-type bonding between p (sulfur) and d_ε (thorium) electrons (see **Fig. 10**). The density of states at the Fermi level was calculated to 1.64 states/eV per mole [91], which is comparable with a value of 1.84 states/eV per mole deduced from the electronic contribution to the specific heat [84].

Neutron inelastic scattering, using time-of-flight spectrometry, gave an optical frequency of 7.8 ± 0.5 THz [72], see also [92], which is comparable with a value of 8.3 ± 0.2 THz [85] obtained by fitting the measured specific heat data with a simple phonon model. The optical phonon frequency, interpreted in terms of a nearest-neighbor force constant gave a value of 6.33×10^5 dyn/cm [72], see also [92]. Results of neutron elastic scattering experiments are given in [73].

The ThVB Auger spectrum of ThS was observed to be very similar to the clean thorium spectrum with only minor differences. It was concluded that the thorium valence band is not appreciably altered by the sulfur atoms in ThS [93].

Electric Resistivity

The electric conductance of ThS (ThS with 85% th. d.) shows a metallic character (n-type) and the electric resistivity a linear increase with temperature (see **Fig. 11**) [70], see also [7, 20, 69, 94 to 97]. Earlier values, given for the room temperature resistivity, are rather different and not well established: ca. $10^{-4} \Omega \cdot$ cm [6], see also [17, 81], or from pressed

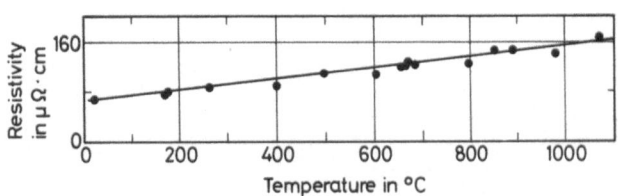

Fig. 11. Temperature dependence of the resistivity of ThS with a density of about 85% th. d. [94].

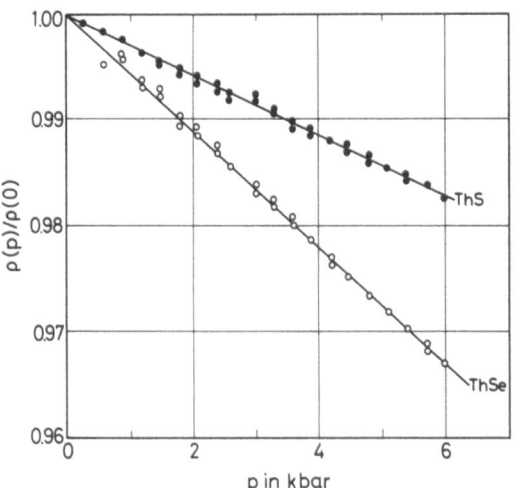

Fig. 12. Relative pressure dependence of the resistivity at 77 K for ThS and ThSe (see p. 89) [100].

powders ca. 0.2 $\Omega \cdot$ cm [98], see also [17, 48]. Further values are 16 to 28 $\mu\Omega \cdot$ cm (ThS with >90% th.d.) [24], see also [17, 26, 27, 36], 16 to 43 $\mu\Omega \cdot$ cm [22], see also [31], 80 $\mu\Omega \cdot$ cm [12], and 85 $\mu\Omega \cdot$ cm (ThS with 85% th.d.) [99].

The electric resistivity of ThS decreases with increasing pressure (see **Fig. 12**) [100], see also [101]. The linear dependence on the pressure, $\varrho(p)$, normalized to zero-pressure resistivity is $d\varrho/\varrho \cdot dp = -2.85 \, (\pm 0.20) \times 10^{-6} \, \text{bar}^{-1}$ [100].

Superconductivity was discovered in ThS at 0.525 K [19]; cf. [20], where no superconductivity was found.

Thermoelectric Power

The Seebeck coefficient, $\alpha = dE/dT$, of thermoelectric power E of ThS (ThS with 85% th.d.) decreases with increasing temperature (see **Fig. 13**) [70], see also [69, 94, 95, 99].

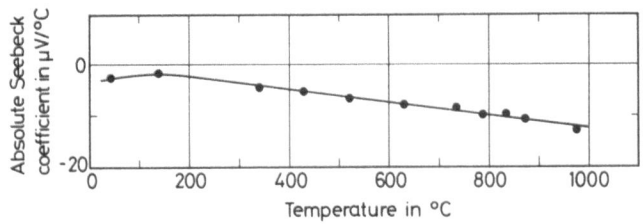

Fig. 13. Temperature dependence of the absolute Seebeck coefficient for sintered ThS [94].

References for 10.1 on pp. 45/8   2*

Hall Coefficient

A value of the Hall coefficient, R_H, for ThS, corrected for porosity, is $R_H = 1.6$ (± 0.2) $\times 10^{-10}$ m^3/Coulomb at 300 K [73].

Thermionic Emission of Electrons

The thermionic emission data of ThS were measured using 0.005 in. ThS coatings on a tungsten wire. The constants from Richardson's law, $I = A \cdot T^2 \cdot \exp(-\varrho/kT)$, are approximately $\varrho = 3.4$ eV and $A = 100$ A $\cdot$ cm$^{-2} \cdot$ K^{-2}. In the vicinity of 1500°C, the emission from ThS is only $^1/_4$ as great as that of ThO$_2$ [79].

10.1.2.6   Magnetic Properties

ThS is reported to be diamagnetic [2, 14], see also [17], or not paramagnetic [6], see also [102]. Otherwise, it was observed that ThS showed a weak paramagnetism in sintered samples as well as in single crystals [102], see also [30]. A value for the room temperature susceptibility is $\chi_{mol} = 50 \times 10^{-6}$ cm^3/mol [7]. For a single crystal, $\chi_g = 0.120 \times 10^{-6}$ cm^3/g was measured, for a sintered sample, 0.085×10^{-6} cm^3/g [102].

10.1.2.7   Optical Properties

The spectral emissivity of ThS, measured using a ThS-coated 0.002 in. molybdenum cylinder, was found to be 0.4 [79].

10.1.2.8   Chemical Reactions

On Heating

ThS is stable up to high temperatures in vacuum as well as under inert gases [6, 22]. The volatility is negligible up to 1900°C in a vacuum of 5×10^{-5} Torr [22]. There was no evidence for any volatility except by decomposition to the gaseous atoms [6]. The heat of atomization is reported to be 151 kcal/g-atom [31]. Loss of weight became significant during sintering of ThS above 2000°C, especially under vacuum conditions. Volatility data observed are summarized in Table 12. For further details, see the chapter on "Thermal Properties" p. 9.

Table 12
Observed Volatility Data for ThS Under Vacuum Conditions.

temperature in °C	loss of weight in wt%/h	Ref.
1900	0.5	[24]
1900	0.6	[36], see also [27]
2000	3	[24], see also [26]
2050	2	[41], see also [22]
2100	8	[24]
2200	moderate	[47]

With the Elements

ThS is reported to be stable towards hydrogen [36]. ThS is quite stable in air at room temperature [6]. Ignition occurs at a temperature of 460 to 510°C; two strong exothermic peaks were observed within this temperature range by DTA measurement, corresponding to the oxidation of ThS to ThOS and ThO_2 [22], see also [10, 17, 24, 27, 36, 40], followed by a weak endothermic peak at 725°C, which could not be explained and might be analogous to that of US at 765°C (see "Uranium" Suppl. Vol. C10, 1984, p. 31) [22].

ThS is quite resistant against alkali metals, alkaline earth metals, aluminium, iron, and titanium even at temperatures above 1500°C unless there is solid solution formation, as observed from melting experiments in ThS crucibles (see Table 13) [42], see also [2, 44, 103, 104]. Disproportionation of ThS occurs with platinum and some other metals of the VIIIth group to Th_2S_3 and a thorium-platinum intermetallic compound [42], see also [44]. It should be noted, that all group IV metal sulfides should be reduced by alkali metals to the metals, as reported by [105], see also [17]. Uranium, cast in ThS crucibles, showed at 1900 to 2000°C only a sulfur up-take of less than 0.15 wt% [106].

Table 13
Experimental Fusions in ThS Crucibles [42].

metal	temperature in °C	time in min	appearance
Ce	1500	15	concave metallic surface with wetting of the sides
Ce	1500	15	sound ingot, wet and stuck to the crucible
Th	1825	6	sound, shiny ingot stuck to the crucible
Mg	900	5	sound ingot which did not stick to the crucible
Al	1500	10	no apparent attack
Fe	1500	10	sound ingot with thin black phase obtained

A Th-ThS eutectic has been observed at 1630°C with a composition of 9.5 at.% sulfur [4], which has to be compared with a value of above 2050°C [6].

With Compounds

If well sintered, ThS is quite resistant to boiling water. Only a very slight surface stain was observed after an exposure of 8 h [6, 22], see also [17, 24, 41]. ThS, if well sintered, dissolves only slowly in dilute oxidizing acids (3 vol% HNO_3 or H_2SO_4) at room temperature [6, 22], see also [41], or $HClO_4$ of a concentration above 1 N [2]. ThS is not attacked by HCl of any strength [22], see also [2, 41].

Halides of alkali or alkaline earth metals can be fused in sintered ThS crucibles without interaction of melt and crucible. Only slow rates of reaction were observed with reactive halides or reactive halide/metal mixtures up to 900°C [42], see also [44].

ThS forms eutectic compositions with ThOS (melting point 1900 to 2000°C) [22], see also [27], and with Th_2S_3 (melting point about 1800°C) [6], see also [2].

Solid solutions are formed with many other monosulfides. Solid solutions with CeS [6, 104], see also [2], and US [6, 22] have been investigated especially.

 References for 10.1 on pp. 45/8

10.1.3 Dithorium Trisulfide, Th_2S_3

10.1.3.1 Formation and Preparation

Reduction of ThS_2

Th_2S_3 was first synthesized by the reaction of ThS_2 and thorium metal; a mixture of ThS_2 and thorium hydride powder in stoichiometric amount was heated up in a first step under reduced pressure to 400 to 600°C to decompose the metal hydride to the metal. At a temperature of 2000 to 2200°C the Th_2S_3 was formed. The reaction was carried out in a molybdenum container; some attack was observed due to alloying of thorium and molybdenum metal [6], see also [9, 17, 64]. The reduction of ThS_2 with ThS was shown to be successful, too. Proper amounts of ThS and ThS_2 powders were pressed into pellets and reacted to give Th_2S_3 at temperatures of 1750 to 1850°C under a vacuum of 10^{-3} to 10^{-5} Torr in molybdenum containers. Such sintered materials were shown to be monophasic without ThS, ThS_2, or any other phases, as indicated by X-ray diffraction measurement [42], see also [6, 43].

Reaction of Thorium Hydride with H_2S

Thorium hydride, ThH_2, prepared from thorium metal and hydrogen at 300°C under reduced pressure, was reacted with H_2S in a stoichiometric amount at 400 to 500°C to give a mixture of metal hydrides and various sulfides. The reaction product was ground in a dry-box, again heated to 500 to 600°C to decompose the hydrides, and then heated at higher temperatures for 30 min to form homogeneous Th_2S_3 [6], see also [2, 17]. The apparatus used is shown in "Uranium" Suppl. Vol. C 10, 1984, p. 3. It is reported to be impossible to remove oxygen impurities from Th_2S_3 [2].

Decomposition of ThS_2

The decomposition of ThS_2, prepared from thorium metal (from thorium hydride powder) and sulfur at 800°C, was carried out in a first step to give Th_7S_{12} by electron beam melting, and then at temperatures above 1650°C under vacuum. The Th_2S_3 obtained was brown in color and had the actual composition of $ThS_{1.40}$, as shown by chemical analysis [8], see also [17, 64].

Synthesis from the Elements

Mixtures of thorium metal powder and sulfur were reacted in a porcelain boat at 800°C for 30 min in a stream of hydrogen. Homogeneous products were obtained ranging from $ThS_{1.22}$ to $ThS_{1.59}$ [14], according to [15, 16], see also [17]. It should be noted that experiments with 2:3 mixtures of thorium metal powder (from decomposition of thorium hydrides) and sulfur, sealed in evacuated silica tubes, failed to reach the equilibrium after heating at 800°C for two weeks; ThS_2 and a second phase of $ThS_{2.5}$ were analyzed by X-ray diffraction measurement [18], see also [17].

Densification of Th_2S_3

Th_2S_3 was sintered by hot-pressing with 350 kg/cm^2 [14], see Table 14. The samples sintered at 1960 to 2000°C consist of two phases, indicating that Th_2S_3 decomposes on sintering [14]. Crucibles of monophasic Th_2S_3 were fabricated by reaction sintering of mixtures of ThS and ThS_2 powders, pressed at 50000 to 100000 psi with a binding agent, and sintered in a vacuum (10^{-3} to 10^{-5} Torr) at 1750 to 1850°C for 30 min in molybdenum containers [42], see also [44].

Table 14
Sintering of Hot-Pressed Th_2S_3 Powders at 350 kg/cm^2 [14].

sintering temperature in °C	time of sintering in min	density in g/cm^3	residual porosity in %
1650	8	4.94	48.8
1750	5	5.72	40.6
1960	8	7.66	20.3
2100	melted	—	—

Enthalpy, Entropy, and Gibbs Free Energy of Formation

First values of the enthalpy of formation of Th_2S_3 were obtained from volatility measurements to be ΔH_f (at 298 K) = -110 ± 15 kcal/S-atom [2], -125 kcal/S-atom [47], and -306 ± 3 kcal/mol [6], see also [14, 17]; for a value of -102 kcal/S-atom, see [48]. A value of ΔH_f (Th_2S_3, 25°C) = -258.6 kcal/mol was calculated from measurements of the heat of solution in 6 M HCl [50], see also [9, 46, 64, 107], or ΔH_f ($ThS_{1.5}$, c, 298 K) = -129.3 ± 2 kcal/mol [13], see also [57, 58]. The heat of formation was derived from combustion calorimetry to be ΔH_f = -119 kcal/S-atom [14], see also [13, 17], or -238 kcal/mol, cited in [46]. An estimated value for the reaction $2Th(c) + 3S_2(g) = Th_2S_3(c)$ at 900°C was calculated by [9] to be -300.5 kcal/mol using the ΔH_f value (at 25°C) of [50], or 279.9 kcal/mol using the value of ΔH_f (at 298 K) cited in [46].

The Gibbs free energy of formation at 1173 K was calculated from emf measurements of solid cells for the reaction $Th(c) + Th_7S_{12}(c) = 4Th_2S_3(c)$ to be ΔG (at 1173 K) = -39.7 ± 2 kcal/g-atom Th [9], see also [46]. From this, ΔH = -47.1 ± 4 kcal/g-atom Th and ΔS = -6.3 ± 3 cal/g-atom were calculated [46].

For more informations, see the section "Thermal Properties", p. 24.

10.1.3.2 Crystallographic Properties

Th_2S_3 is orthorhombic (Sb_2S_3 type) with four molecules per unit cell; the space group is Pbnm-D_{2h}^{16} (No. 62) [5]. The lattice parameters were determined from a small single crystal to be (in kX) a = 10.97 ± 0.05, b = 10.83 ± 0.05, c = 3.95 ± 0.05 [5], also cited in [2, 3, 6, 12, 14, 18, 64, 65], or (in Å) a = 10.99, b = 10.85, c = 3.96 [18] from data of [5], also cited in [13, 17, 60, 67, 68]. The theoretical density was calculated from these values to be 7.87 g/cm^3 [5] or 7.88 g/cm^3 [6], see also [81].

The atomic coordinates are $\pm$(x, y, $^1/_4$); ($^1/_2-$x, y$+^1/_2$, $^1/_4$) (origin in an inversion center) [5]:

Th(1):   x = 0.314 ± 0.003;   y = -0.022 ± 0.003
Th(2):   x = 0.519 ± 0.003;   y = 0.300 ± 0.003
S(1):   x = 0.878;   y =   0.053
S(2):   x = 0.561;   y =   0.871
S(3):   x = 0.206;   y =   0.230

                                References for 10.1 on pp. 45/8

The thorium-sulfur distances (in Å) are [5]:

Th(1)-2 S(1) = 2.91	Th(2)-1 S(1) = 2.96	Mean value: Th-S = 2.90 Å.
Th(1)-2 S(2) = 2.91	Th(2)-2 S(1) = 2.97	
Th(1)-1 S(2) = 2.94	Th(2)-2 S(2) = 2.84	
Th(1)-1 S(3) = 2.83	Th(2)-2 S(3) = 2.86	
Th(1)-1 S(3) = 2.86	(Th(2)-1 S(3) = 3.56)	

Th_2S_3 is described as a sharply defined chemical composition [5], whereas a homogeneity range extending from $ThS_{1.22}$ to $ThS_{1.59}$ was found by [14]; X-ray diffraction patterns are given for this range [14].

The oxidation state of the thorium ions in Th_2S_3 is $+4$, predicted theoretically on the basis of a 5f contraction in an actinide series. There are no 5f electrons in Th_2S_3. The predicted interatomic distance is Th-S = 3.02 Å [108], compared with the measured mean value of Th-S distance, 2.90 Å [5].

A value for the entropy of the crystal lattice was calculated to be $S = 158.7$ cal $\cdot$ mol^{-1} $\cdot$ K^{-1} [109].

10.1.3.3 Mechanical Properties

The theoretical density of Th_2S_3 was calculated from X-ray measurements to be 7.87 to 7.88 g/cm^3 [5, 6], see p. 23. A densification of up to 7.66 g/cm^3 was achieved by hot-pressing [14], see p. 22.

Measured values of the microhardness of different samples are reported: powder particles, 171.1 ± 29.91 kg/mm^2 (5 g load); sintered specimen, 539 ± 45 kg/mm^2 (30 g load); and sintered specimen after annealing, 227 ± 28 kg/mm^2 (20 g load) [14].

10.1.3.4 Thermal Properties

Th_2S_3 is nonvolatile at 1700 °C [6], see also [47] and melted with no appreciable volatility [2], see also [14].

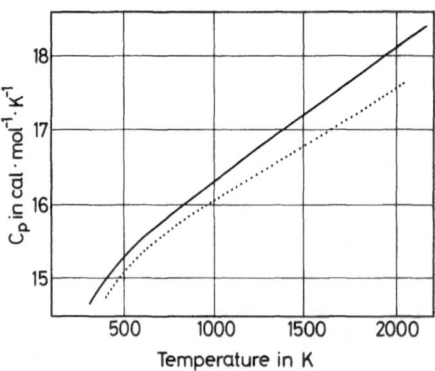

Fig. 14. Heat capacity of $ThS_{1.50}$(c) [13]; ——— values estimated by [110]; ······ values from $\frac{1}{2}[C_p(ThS) + C_p(ThS_2)]$ [13].

The melting point of Th_2S_3 was determined to be $1950 \pm 50\,°C$ [6], see also [2, 11, 12, 17, 47, 64, 65, 80 to 82] or about $2300\,°C$ [14]. Th_2S_3 samples were observed to decompose on sintering within the temperature range of 1960 to $2000\,°C$ [14].

Heat capacity values of Th_2S_3 were estimated over a temperature range of 298 to 2270 K [110]; these values should be compared to values calculated from $\frac{1}{2}\,[C_p(ThS) + C_p(ThS_2)]$ [13], shown in **Fig. 14**. The estimated values of the heat capacity and the derived values for the other thermodynamic functions are summarized in Table 15, p. 26, see [13].

The standard entropy $S°(ThS_{1.5}, c, 298) = 20 \pm 2$ cal $\cdot$ mol^{-1} $\cdot$ K^{-1} (estimated) [86], see also [46], compares to $S°(ThS_{1.5}, c, 298) = 21 \pm 1.5$ cal $\cdot$ mol^{-1} $\cdot$ K^{-1} [57], see also [13], calculated from emf measurements of solid-electrochemical cells [9]. Another value of 34.7 cal $\cdot$ mol^{-1} $\cdot$ K^{-1} is tabulated in [109].

10.1.3.5 Electrical Properties

Th_2S_3 is reported to be a semiconductor (n-type) with a room temperature resistivity of 10^{-3} to 10^{-4} $\Omega \cdot$ cm [6], also cited in [2, 17, 81], see also [101], $10.0\,\Omega \cdot$ cm (powders pressed at 25000 psi) [98], also cited in [17], or $1.0\,\Omega \cdot$ cm [48].

The thermoelectric power of Th_2S_3 was measured at pressed powders (100000 psi) to be 0.1 mV/°C [98], see also [48].

10.1.3.6 Magnetic Properties

Th_2S_3 has been reported to be diamagnetic [14], see also [17], or not paramagnetic [6]. However, a weak temperature-independent paramagnetism was observed from susceptibility measurements, performed on sintered samples at 77 to 300 K, with a room temperature value $\chi_g = 0.280 \times 10^{-6}$ cm^3/g [102]. This agrees with an earlier observation, from which Th_2S_3 was shown to be paramagnetic with evidence of slightly ferromagnetic behavior even though iron-free [2].

10.1.3.7 Chemical Reactions

On Heating

The brown Th_2S_3 is nonvolatile at temperatures up to $1700\,°C$ [6], see also [47]. It melts at 1900 to $2000\,°C$ without appreciable volatility [2], see also [64]. However, decomposition was observed at 1960 to $2000\,°C$ on sintering [14].

For further details see the section on "Thermal Properties", p. 24.

With the Elements

Th_2S_3 is quite stable in air at room temperature [6], see also [64].

Th_2S_3 is reduced by reactive metals to ThS and the corresponding metal sulfides, as observed from melting experiments in Th_2S_3 crucibles [42], see also [2, 44, 105]. Alkali and alkaline earth metals formed separate phases, beryllium a volatile sulfide, magnesium and barium attacked the crucibles at even moderate temperatures, while lanthanum, cerium, other rare earth metals, thorium, and uranium formed MeS-Th_2S_3 solid solutions [42], see also Table 16.

                    References for 10.1 on pp. 45/8

Table 15
Thermodynamic Functions of Dithorium Trisulfide, $Th_2S_3(c)$ [13].
See note in Table 9 on p. 15.

T in K	C_p° in cal·mol^{-1}·K^{-1}	S° in cal·mol^{-1}·K^{-1}	$-(G_T^\circ - H_{298}^\circ)/T$ in cal·mol^{-1}·K^{-1}	$H_T^\circ - H_{298}^\circ$ in cal/mol	ΔH_f° in cal/mol	ΔG_f° in cal/mol	log K_p
298	14.606	21.500	21.500	0	−129500	−128680	94.323
300	14.615	21.590	21.500	27	−129500	−128675	93.738
400	15.013	25.852	22.079	1509	−130330	−128363	70.133
500	15.300	29.234	23.183	3025	−130911	−127806	55.863
600	15.530	32.044	24.432	4567	−131361	−127135	46.308
700	15.743	34.454	25.696	6131	−131725	−126406	39.465
800	15.944	36.570	26.925	7715	−151526	−127424	34.810
900	16.140	38.459	28.104	9320	−151348	−124422	30.213
1000	16.330	40.169	29.226	10943	−151176	−121440	26.540
1100	16.518	41.734	30.293	12585	−151012	−118474	23.538
1200	16.703	43.180	31.308	14247	−150852	−115523	21.039
1300	16.888	44.524	32.273	15926	−150700	−112586	18.927
1400	17.072	45.782	33.194	17624	−150553	−109659	17.118
1500	17.255	46.966	34.073	19340	−150413	−106742	15.552
1600	17.437	48.086	34.914	21075	−150278	−103835	14.183
1700	17.619	49.148	35.720	22828	−150965	−100902	12.972
1800	17.801	50.161	36.494	24599	−150782	−97963	11.894
1900	17.982	51.128	37.239	26388	−150614	−95033	10.931
2000	18.163	52.055	37.957	28195	−150461	−92113	10.066
2100	18.345	52.945	38.650	30021	−153727	−89071	9.270
2200	18.525	53.803	39.319	31864	−153712	−85993	8.542

Table 16
Experimental Metal Fusions in Th_2S_3 Crucibles [42].

metal	temperature in °C	time in min	appearance
Ce	1500	15	flat, smooth metallic surface; ingot stuck to crucible
Ce	1500	15	concave surface; wetting crucible sides; ingot stuck to crucible
Th	1825	6	crucible broke away cleanly leaving thin coating of ThS on ingot

With Compounds

Well sintered Th_2S_3 is quite stable against water even if boiling [6], see also [64].

Th_2S_3 is only slowly dissolved in dilute acids at room temperature [6], see also [105]. Dissolution occurs readily in > 1 N HCl and $HClO_4$ [2], see also [50, 64, 107]. A slow but steady reaction was observed in 2 N H_2SO_4 at room temperature with liberation of H_2S [105]. Ignition occurs if Th_2S_3 is attacked by concentrated HNO_3 [105], see also [6].

With ThS, Th_2S_3 forms a eutectic composition which melts at about 1800°C [6] or about 1900°C [11], see also [2].

10.1.4   Heptathorium Dodecasulfide, Th_7S_{12}

The homogeneity range of Th_7S_{12} extends from $ThS_{1.71}$ to $ThS_{1.76}$ [3, 6]. The formula of this compound might thus be Th_3S_5, Th_4S_7, or Th_7S_{12}. The actual formula of this compound was determined by X-ray measurements of a single crystal to be Th_7S_{12} [3].

10.1.4.1   Formation and Preparation

Th_7S_{12} was first prepared by thermal decomposition of ThS_2 and by the reaction of thorium hydride with H_2S [6], which methods are still used for preparation. Besides this, other preparative techniques were developed.

Decomposition of ThS_2

The thermal decomposition of ThS_2, carried out in molybdenum containers under vacuum at temperatures above the melting point of ThS_2 (>1950°C), rapidly leads to the formation of Th_7S_{12} compounds with compositions ranging from $ThS_{1.71}$ to $ThS_{1.76}$ [6], see also [2, 3, 14, 17]. Th_7S_{12} with an S:Th ratio of 1.75 was obtained after 3 h at 1900 to 2000°C under a vacuum of 10^{-5} Torr [16], see also [14]. $ThS_{1.76}$ was obtained from ThS_2 by means of electron beam melting [8].

Reaction of Thorium or Thorium Hydride with H_2S

Thorium hydride, prepared from thorium metal at temperatures of up to 350°C, was reacted with H_2S in stoichiometric amount at 400 to 500°C to form a mixture of thorium hydrides and various sulfides. The reaction product, ground and mixed in an inert box to prevent oxidation, was then heated up to 500 to 600°C to decompose the hydrides, and homogenized at higher temperatures forming Th_7S_{12} [6], see also [2]. The reaction of thorium hydride with H_2S was carried out in an apparatus as shown schematically in "Uranium" Suppl. Vol. C10, 1984, p. 3; a molybdenum container was used for the further homogenization steps [6]. It is reported to be impossible to remove oxygen impurities from Th_7S_{12} [2].

$Th_{7-x}S_{12}$ was prepared by the reaction of thorium powder (2.5 µm grain size) with H_2S (2 to 110 Torr H_2S) at a temperature of 440 to 540°C. The activation energy of this reaction was determined to be 32 ± 2 kcal/mol [111].

Attempted Synthesis from the Elements

The reaction of thorium metal (from thorium hydride) and sulfur to form Th_7S_{12} failed after two weeks of heating the reaction mixture at 800°C in evacuated and sealed silica tubes. The reaction product was ground, re-evacuated, and again heated to 800°C for seven days, and determined to consist of ThS_2 and an unidentified second phase [18], see also [15 to 17].

Other Methods

Th_7S_{12} was obtained from the reaction of ThS_2 with thorium metal (from thorium hydride) at 1600°C within 1 h. The reaction product was crushed, pelletized, and re-heated at 1600°C for homogenization [9].

ThO_2 was reacted with H_2S in the absence of carbon. The reaction was carried out in a porcelain boat, placed in a silica tube, under a stream of H_2S. The reaction product was ThOS, when heating up to 1100°C. $ThS_{1.7}$ was obtained at temperatures of 1200 to 1300°C. The products contained 99.2% thorium sulfide, 0.1% free sulfur, and 0.8% oxygen (calculated) as determined by chemical analysis. The presence of the Th_7S_{12} structure was determined by X-ray diffraction measurement [112], see also [113, 114].

Densification

Crucibles of monophasic Th_7S_{12} were fabricated by reaction sintering of mixtures of ThS and ThS_2 powders, pressed at 50000 to 100000 psi with a binding agent, and sintered in vacuum (10^{-3} to 10^{-5} Torr) at 1750 to 1850°C for 30 min in molybdenum containers [42], see also [43, 44, 115].

Enthalpy, Entropy, and Gibbs Free Energy of Formation

Estimated values of the enthalpy of formation of Th_7S_{12} at room temperature are $\Delta H_f = -114$ kcal/S-atom [47], -665 ± 35 kcal/mol [6], see also [13, 14, 17, 46], -95 kcal/S-atom [48], -100 ± 10 kcal/mol ($ThS_{1.71}$) [116], see also [13], -142 ± 10 kcal/mol ($ThS_{1.71}$) [57], see also [13], and -141.3 kcal/mol ($ThS_{1.71}$) [58], see also [13]. The enthalpy of formation derived from measurements with solid emf cells was calculated to be ΔH_f° (at 1173 K) $= -89.4$ kcal/ g-atom S [9] or -96.2 kcal/g-atom S [9], see also [49], using the enthalpy of formation of Th_2S_3 from [50] or the enthalpy of vaporization from [51], respectively, see also [46]. A value of ΔH_f° ($ThS_{1.71}$, c, 298) $= -141$ kcal/mol or -142 ± 3 kcal/mol was calculated for the reaction Th(c) $+ Th_7S_{12}(c) = 4Th_2S_3(c)$ from the already mentioned galvanic cell measurements [52], see also [13]. A value for the entropy of formation has been estimated to be ΔS° (at 298 K) $= -17.55$ cal $\cdot$ (g-atom S)$^{-1} \cdot$ K^{-1} [59]. The Gibbs free energy of formation at 1173 K was calculated from emf measurements of solid electrochemical cells for the reaction $Th(c) + 6 ThS_2(c) = Th_7S_{12}(c)$ to be ΔG (at 1173 K) $= -95.2$ kcal/g-atom Th [9], see also [46]. From this, a value of -102 kcal/g-atom Th was estimated [46]. Other estimated values for ΔS_f° and ΔG_f° are given in [59].

10.1.4.2 Crystallographic Properties

Th_7S_{12} is hexagonal (Th_7S_{12} type) with one molecule per unit cell; the space group is $P6_3/m$-C_{6h}^2 (No. 192) [3], see also [60]. Th_7S_{12} has a wide homogeneity range with S:Th ratios of 1.71 (stoichiometric compound) to 1.76 resulting in the formula $Th_{7-x}S_{12}$ with $0 \leqslant x \leqslant 0.2$ [3]. The lattice parameters were determined from small needle-like single crystals. For an atomic ratio S/Th = 1.71, a = 11.041 $\pm$ 0.001 kX, c = 3.983 $\pm$ 0.001 kX [3], see also [6, 12, 14, 64, 65, 117], or a = 11.063 Å, c = 3.991 Å [18] from [3], see also [13, 17, 67]. For S/Th = 1.76, a = 11.064 $\pm$ 0.002 kX, c = 4.002 $\pm$ 0.002 kX [3], see also [117], or a = 11.086 Å, c = 4.010 Å, c/a = 0.3617 [60] from [3]. The lattice constants of a Th_7S_{12} powder (81.5 to 81.3% Th, 17.7 to 17.9% S, 0.8% O) are determined to be a = 11.045 kX, c = 3.984 kX [112], see also [113]. From this, the theoretical densities are calculated to range from 7.85 (S/Th = 1.71) to 7.65 g/cm^3 (S/Th = 1.76) [2, 3], see also [64, 65, 117]. A value of 7.88 g/cm^3 is given in [83].

Each of the Th(1) atoms is coordinated to nine S(2) atoms, while each Th(2) atom is coordinated to five S(1) and to three S(2) atoms. The atomic positions of the $P6_3/m$ structure are 1 Th(1) in $\pm$ (0, 0, $^1/_4$), 6 Th(2), 6 S(1), 6 S(2) in $\pm$ (x, y, $^1/_4$); ($\bar{y}$, x$-$y, $^1/_4$); (y$-$x, $\bar{x}$, $^1/_4$), with parameters Th(2): x = 0.153 $\pm$ 0.002, y = $-0.283 \pm$ 0.002, S(1): x = 0.514 $\pm$ 0.010, y = 0.375 $\pm$ 0.010, S(2): x = 0.235 $\pm$ 0.010, y = 0 $\pm$ 0.010 (mean values) [3], see also [60, 117].

The determined space group of $P6_3/m$ does not strictly apply to this structure. Since there is only one Th(1) atom but two equivalent a sites ((0, 0, $^1/_4$); (0, 0, $^3/_4$)) the Th(1)-S(2) distances are different depending on whether the (0, 0, $^1/_4$) or the (0, 0, $^3/_4$) position is occupied or vacant. From this, the S(2) atoms are expected to be shifted from this mean position (see **Fig. 15**). Thus, the interatomic distances are Th(1)-3 S(2) = 2.82 Å, Th(2)-2 S(1) = 2.95 Å, Th(2)-2 S(2) = 2.85 Å, Th(1)-6 S(2) = 3.09 Å, Th(2)-3 S(1) = 2.98 Å, Th(2)-1 S(2) = 3.00 Å [3], see also [117].

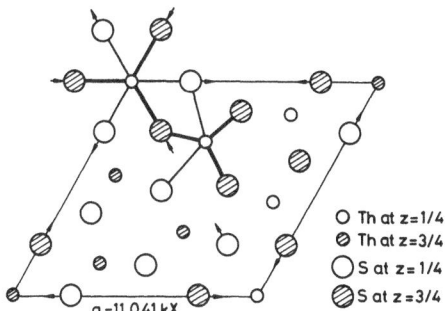

O Th at z=1/4
⊘ Th at z=3/4
◖ S at z=1/4
◕ S at z=3/4

a=11.041 kX

Fig. 15. The Th_7S_{12} structure viewed along a sixfold axis. Two of the Th(1) atoms shown in the figure are placed at z = $^1/_4$ and two at z = $^3/_4$. Arrows indicate displacements of S(2) atoms from their mean sites [3].

References for 10.1 on pp. 45/8

10.1.4.3 Mechanical Properties

The theoretical density of Th_7S_{12} was calculated from X-ray measurements to be 7.85 to 7.65 g/cm^3 due to the homogeneity range of S/Th = 1.71 to 1.76 [3], see p. 29.

The microhardness of Th_7S_{12} (ThS$_{1.7}$) is reported to be 433 ± 30 kg/mm^2 (10 g load). The sample was prepared by the reaction of ThO_2 with H_2S at a temperature of 1200 to 1300°C (see p. 28) [112], see also [113].

10.1.4.4 Thermal Properties

Th_7S_{12} is nonvolatile at 1700°C [6], see also [2].

The melting point of Th_7S_{12} was determined to be 1700 ± 30°C [6], see also [2, 12, 14, 17, 65, 80, 82, 118]. Higher melting points are reported to be 1800 to 1900°C [11], 1950°C [81], and 2000°C [98].

Values of the entropy of Th_7S_{12} are estimated to be S°(ThS$_{1.71}$, c, 298) = 21 ± 2 cal · mol^{-1} · K^{-1} [86], see also [46], 21.9 ± 2 cal · mol^{-1} · K^{-1} [52], based on emf measurements [9] for the reaction Th(c) + Th_7S_{12}(c) = 4 Th_2S_3(c), and 22 ± 2 cal · mol^{-1} · K^{-1} [57], calculated from the values of ThS$_{1.50}$ and ThS$_2$.

10.1.4.5 Electrical Properties

Th_7S_{12} is a semiconductor (n-type) with a room temperature resistivity of 10^{-3} to 10^{-4} Ω · cm (under 25000 psi) [6], also cited in [81, 98]. This value is considered as very low, as it approaches the value for pure Th metal [98]. Other values are 2.5×10^4 Ω · cm (powders pressed at 25 000 psi) [98], also cited in [17, 48], or 3.6×10^7 to 7×10^7 Ω · cm [112], also cited in [48].

The thermoelectric power of Th_7S_{12} was measured on pressed powders (100000 psi) to be 0.2 mV/°C [98], also cited in [48].

10.1.4.6 Magnetic Properties

Th_7S_{12} is reported to be not paramagnetic [6] or to be diamagnetic [14], see also [17]. However, a weak temperature-independent paramagnetism was observed from susceptibility measurements, performed on sintered samples at 77 to 300 K, with a room temperature value of $\chi = 0.230 \times 10^{-6}$ cm^3/g [102].

10.1.4.7 Chemical Reactions

With Elements

Th_7S_{12} is quite stable in air at room temperature [6]. The oxidation in oxygen atmosphere starts at 500°C and is completed at 1300°C with the formation of ThO_2 (see Table 17) [114].

Table 17
Oxidation of $ThS_{1.7}$ in an Oxygen Atmosphere.
The sulfur content of the original sulfide was 17.8% [114].

temperature in °C	S oxidized (in wt%) after 10 to 60 min						S content of the oxidized sulfide after 1 h, in wt%
	10	20	30	40	50	60	
300	0	0	0	0	0	0	17.8
400	0	0	0	0	0	0	17.8
500	0	0	0	0	0	0	17.8
600	0.8	1.8	2.0	2.2	2.8	3.0	14.8
700	2.1	4.1	5.25	6.0	6.2	6.8	11.0
800	3.2	7.0	10.25	12.2	11.0	13.8	4.0
900	4.2	8.9	12.1	13.2	14.0	14.7	3.1
1000	5.85	11.1	13.9	14.2	14.92	15.6	2.2
1100	8.0	13.15	14.95	15.2	15.9	16.5	1.3
1200	10.32	14.65	15.25	16.0	16.2	17.0	0.8
1250	12.15	15.5	16.0	16.3	17.0	17.3	0.5
1300	15.0	16.15	16.9	17.2	17.7	17.8	0

Th_7S_{12} is reduced by reactive metals to ThS and the corresponding metal sulfides, as observed from melting experiments in sintered Th_7S_{12} crucibles [42], see also [2, 44, 105]. Alkali and alkaline earth metals formed separate phases, beryllium a volatile sulfide, and barium attacked the crucibles at even moderate temperatures, while lanthanum, cerium, other rare earth metals, thorium, and uranium formed solid solutions (see Table 18).

Table 18
Experimental Metal Fusions in Th_7S_{12} Crucibles [42].
Written as "Th_4S_7" in the original table; see p. 27.

metal	temperature in °C	time in min	appearance
U	1300	5	bright metallic ingot; removable from crucible by tapping
U	1475	30	concave surface on ingot; removable from crucible by tapping
Ce	1500	15	sound ingot sticking to crucible
Ce	1500	15	sound ingot sticking to crucible; thin layer of crucible on ingot

With Compounds

Well sintered Th_7S_{12} is quite stable against water, even if boiling [6, 119]. It is only slowly dissolved in dilute acids at room temperature [6]. Dissolution readily occurs in >1 N HCl and $HClO_4$ [2]. The values listed in Table 19, p. 32, were observed after dissolution in hot acids and NaOH.

With ThS_2, Th_7S_{12} forms a eutectic composition which melts at 1765 ± 25°C [2], at 1800 to 1850°C [11].

Table 19
Dissolution of Th_7S_{12} in Hot Acids and in Sodium Hydroxide after an Exposure of 30 min [119].

medium of dissolution	temperature in °C	content of sulfide in wt% of the residue
H_2SO_4 (1:1)	136	hydrolysis
HCl (D = 1.19)	112	8.5
H_3PO_4 (1:1)	110	20.5
$H_2C_2O_4$ (saturated)	100	24.2
CH_3COOH (glacial)	100	18.6
NaOH (20%)	100	20.5

10.1.5   Thorium Disulfide, ThS_2

10.1.5.1   Formation and Preparation

The formation of a thorium (di-)sulfide was observed for the first time already in 1829 [120]. Several preparative methods were reported in the following decades but the products obtained were mostly not well defined or characterized. Special information is given in "Thorium" 1955, pp. 277/83, see also [64]. The earliest study of a well defined ThS_2 was presented in 1932/33 [121, 122]. Lastly, three preparative methods were developed, which are well established: 1. reaction of thorium metal with elementary sulfur, 2. reaction of thorium or thorium hydride with H_2S, and 3. reaction of ThO_2 with H_2S in the presence of carbon.

Synthesis from the Elements

Thorium filaments were reacted with sulfur in stoichiometric amounts at 600°C for 18 h and then at 780 to 800°C for 100 h using a special pressure-synthesis technique. The ThS_2 obtained was violet-brown in color [1], see also [14, 15, 64]. A polysulfide, Th_2S_5, was prepared in a first step, when the reaction was carried out with sulfur in excess and sealed in a silica tube at 400 to 420°C for 2 to 3 weeks to ensure complete reaction and reasonable crystal growth. The excess sulfur was sublimed at 250 to 300°C. $ThS_{2.00}$, purple-brown in color, was obtained from this Th_2S_5 by thermal decomposition in vacuum at 900°C after several hours [18], see also [17]. Purple colored ThS_2 was obtained from the reaction of thorium metal powder, prepared from decomposition of thorium hydride, with sulfur at 800°C [8].

Reaction of Thorium or Thorium Hydride with H_2S

Thorium hydride, prepared from thorium metal at temperatures of up to 350°C, was reacted with H_2S in stoichiometric amounts at 400 to 500°C to form ThS_2. The reaction was carried out in an apparatus, as shown schematically in "Uranium" Suppl. Vol. C10, 1984, p. 3. It was observed that no thorium hydride or lower sulfides remained when the reaction was completed [6]. But a purer product was obtained by the reaction of ThO_2 with H_2S in the presence of carbon [6], see also [2, 9, 64].

A dark brown ThS_2 was obtained from the reaction of thorium metal powder, prepared by decomposition of thorium hydride at 350 to 400°C, with H_2S at a temperature of 300 to 400°C. Small amounts of thorium metal impurities were detected by X-ray diffraction measurement [124], see also [17, 64, 125].

Reaction of ThO$_2$ with H$_2$S in the Presence of Carbon

Thorium dioxide reacts with H$_2$S in the presence of carbon (i. e., ThO$_2$ placed in a graphite crucible) to form ThS$_2$. The reaction was carried out in a first step at 1200 °C for 1 h and then, after completion of the reaction at 1600 °C for 30 min, a uniform product, black in color (79.09 to 78.87 wt% Th, 21.65 to 21.3 wt% S), was obtained [121], see also [17, 64, 122]. The reaction is reported to run fairly rapidly at a temperature of 1200 to 1300 °C, whereas ThOS is formed as an intermediate product with liberation of water and later CO. The formation of CS$_2$, formed as a further intermediate by the reaction of H$_2$S with the carbon crucible, is important in the reaction with ThOS to produce ThS$_2$. Last traces of oxygen are removed at 1400 to 1500 °C. The ThS$_2$ formed is reported to be purple in color [6], see also [2, 9, 64, 126]. A further investigation of this reaction led to ThS$_{1.97}$, which was brown in color. The observed intermediate products, based on chemical analysis, are summarized in Table 20 [112], see also [28, 113, 114].

Table 20
Chemical and Phase Composition of the Products of Interaction of ThO$_2$ and H$_2$S in the Presence of Carbon [112].

temperature in °C	time in h	chemical composition in wt%				phase composition
		Th	S	Th + S	oxygen (by difference)	
800	1	86.8	5.02	91.82	8.18	ThO$_2$ + ThOS
800	2	86.1	8.45	94.55	5.45	ThOS
900	1	83.3	9.65	92.95	7.05	ThOS
900	2	83.2	9.86	93.06	6.94	ThOS
1000	1	83.1	11.5	94.6	5.4	ThOS
1000	2	83.2	11.5	94.7	5.3	ThOS
1100	1	81.6	15.5	97.1	2.9	ThS$_{1.8}$ + ThOS
1100	2	81.05	15.8	96.85	3.15	ThS$_{1.8}$ + ThOS
1200	1	79.2	19.2	98.4	1.6	ThS$_{1.9}$ + ThOS
1200	2	79.0	19.5	98.5	1.5	ThS$_{1.9}$ + ThOS
1300	1	79.11	19.9	99.01	0.99	ThS$_2$
1300	2	79.11	20.5	99.71	0.39	ThS$_2$

Other Methods

A purple colored ThS$_2$ was produced by the decomposition of the Th^{IV} complex of tetramethylthiuramdisulfide, Th(NO$_3$)$_4$ · C$_6$H$_{12}$N$_2$S$_4$ [127]. The intermediate reaction products, based on thermogravimetric measurements, are summarized in Table 21, p. 34. The neutral organic ligands involved are

C$_6$H$_{12}$N$_2$S$_4$ = tetramethylthiuramdisulfide, (CH$_3$)$_2$N-C-S-S-C-N(CH$_3$)$_2$;
$\qquad\qquad\qquad\qquad\qquad\qquad\qquad$ ‖ $\quad$ ‖
$\qquad\qquad\qquad\qquad\qquad\qquad\qquad$ S $\quad$ S

C$_6$H$_{12}$N$_2$S$_3$ = tetramethylthiurammonosulfide, (CH$_3$)$_2$N-C-S-C-N(CH$_3$)$_2$;
$\qquad\qquad\qquad\qquad\qquad\qquad\qquad\quad$ ‖ $\quad$ ‖
$\qquad\qquad\qquad\qquad\qquad\qquad\qquad\quad$ S $\quad$ S

C$_5$H$_{12}$N$_2$S $\;$ = tetramethylthiocarbamide, (CH$_3$)$_2$N-C-N(CH$_3$)$_2$.
$\qquad\qquad\qquad\qquad\qquad\qquad\qquad$ ‖
$\qquad\qquad\qquad\qquad\qquad\qquad\qquad$ S

Table 21
Thermal Decomposition (TG) Data for the $Th(NO_3)_4$ Complex with Tetramethylthiuramdisulfide [127].

complex	decomposition temperature in °C		decomposition product	weight loss in %	
	initial	final		found	calc.
$Th(NO_3)_4 \cdot C_6H_{12}N_2S_4$	130	155	$Th(NO_3)_4 \cdot C_6H_{12}N_2S_3$	5.32	4.44
	230	310	$Th(NO_3)_4 \cdot C_5H_{12}N_2S$	16.32	15.00
	410	570	ThS_2	60.10	58.88

ThS_2 was also prepared by extracting Th^{IV} thiocyanate into moist 1-pentanol. ThS_2, purple-brown in color, was isolated following distillation and calcination of the residue under a stream of H_2S (see the flowsheet of **Fig. 16**) [128], see also [129].

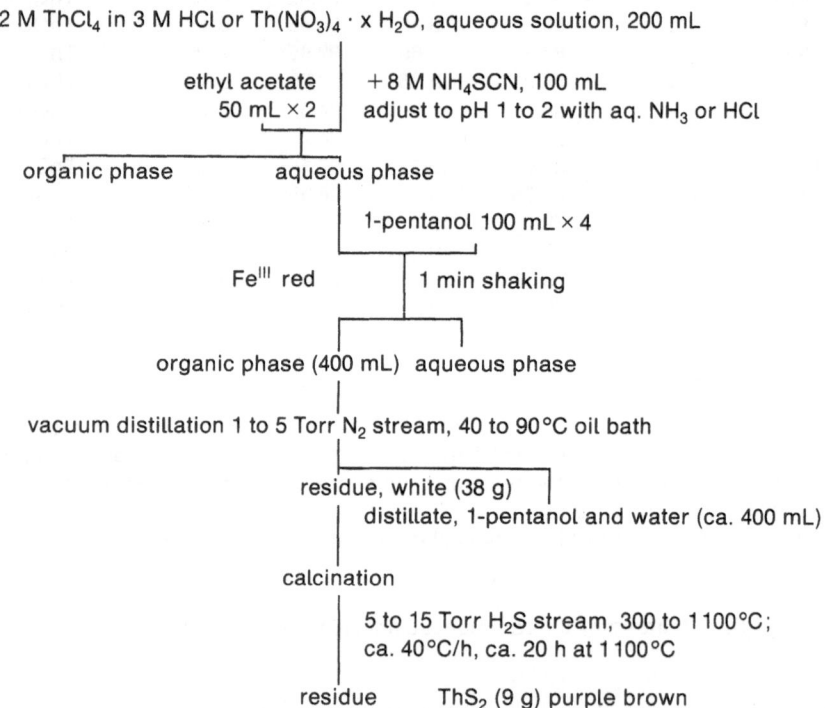

0.2 M $ThCl_4$ in 3 M HCl or $Th(NO_3)_4 \cdot x\ H_2O$, aqueous solution, 200 mL

ethyl acetate
50 mL × 2   + 8 M NH_4SCN, 100 mL
                  adjust to pH 1 to 2 with aq. NH_3 or HCl

organic phase   aqueous phase

1-pentanol 100 mL × 4

Fe^{III} red   1 min shaking

organic phase (400 mL) aqueous phase

vacuum distillation 1 to 5 Torr N_2 stream, 40 to 90 °C oil bath

residue, white (38 g)
                  distillate, 1-pentanol and water (ca. 400 mL)

calcination

5 to 15 Torr H_2S stream, 300 to 1100 °C;
ca. 40 °C/h, ca. 20 h at 1100 °C

residue ThS_2 (9 g) purple brown

Fig. 16. Preparation of ThS_2 from thorium(IV) thiocyanate [128].

Densification

ThS$_2$ crucibles were fabricated by pressing the powders at 50000 to 100000 psi with a binding agent and sintering at 1600 to 1900°C for 30 min in a vacuum (10^{-3} to 10^{-5} Torr) in molybdenum containers [42], see also [44].

Single Crystals

Single crystals of ThS$_2$ were obtained by the same procedure as used for ThOS single crystals [130] (see p. 49) with a transport reaction using iodine as a transporting agent [131, 132].

Enthalpy, Entropy, and Gibbs Free Energy of Formation

Values for the heat of formation, ΔH_f°, at room temperature were estimated to be -170 ± 20 kcal/mol (at 298 K) [6], see also [2, 13, 14, 17, 46], -106 kcal/g-atom S (at 290 K) [47], -151 ± 20 kcal/mol [133], see also [13], or -189 ± 20 kcal/mol [134], see also [13]. Values of ΔH_f° (at 298 K) $= -176 \pm 5$ kcal/mol [135] or -174 ± 5 kcal/mol [135] were calculated evaluating the reaction ThO$_2$(c) + 2 H$_2$S(g) = ThS$_2$(c) + 2 H$_2$O(g), see also [13], and -163.5 ± 5 kcal/mol [135] or -170 ± 5 kcal/mol [135] evaluating the reaction ThO$_2$(c) + CS$_2$(g) = ThS$_2$(c) + CO$_2$(g), see also [13]. A value of -195 ± 6 kcal/mol was derived from combustion calorimetry [136], see also [13, 17]. The enthalpy of formation, derived from emf measurements with solid cells, was calculated to be ΔH_f° (at 1173 K) $= -80.9$ kcal/g-atom S [9] or -87.7 kcal/g-atom S [9], see also [46, 49] using the enthalpy of formation of Th$_2$S$_3$ from [50] or the entropy of vaporization of ThS from [51], respectively. A value of ΔH_f° (at 298 K) $= -149.3 \pm 5$ kcal/mol [52], see also [13] or -150 ± 10 kcal/mol [57], see also [13] was later calculated evaluating the reaction Th(c) + 6 ThS$_2$(c) = Th$_7$S$_{12}$(c) from the above-mentioned galvanic cell measurements.

The standard entropy of formation, ΔS_f° (at 298.15 K) $= -5.0 \pm 0.3$ cal $\cdot$ mol^{-1} $\cdot$ K^{-1}, was derived from measurements of the heat capacity in the temperature range of 51 to 298 K [137]. This value was used to calculate the Gibbs free energy of formation ΔG_f° (ThS$_2$, c, 298.15) $= -137.5$ kcal/mol (from metal and rhombic sulfur) or -156.8 kcal/mol (from metal and S$_2$(g)) [137]. Further estimated values are given in [59].

10.1.5.2 Crystallographic Properties

ThS$_2$ is orthorhombic (PbCl$_2$ type) with four molecules per unit cell; the space group is Pmnb-D$_{2h}^{16}$ (No. 62) [5], see also [60, 131]. The lattice parameters, measured by X-ray diffraction on powdered samples [5, 18, 112] or a single crystal [131], were determined to be a = 4.259 ± 0.002, b = 7.249 ± 0.003, c = 8.600 ± 0.003 kX [5], see also [6, 14, 60, 64, 65, 131]; or given in Å: a = 4.268, b = 7.264, c = 8.617 [18] from [5], see also [13, 17, 26, 60, 67, 68, 112]. a = 4.283, b = 7.275, c = 8.617 Å [18] in good agreement with [5], see also [12, 13, 17, 67]. a = 4.267, b = 7.262, c = 8.617 Å [112], see also [113].

From this, the theoretical density was calculated to be 7.36 g/cm^3 [5], see also [14, 64, 65, 112, 113], compared to a measured pycnometric density of 7.30 g/cm^3 [1], see also [14], or 7.31 g/cm^3 [112], see also [113].

References for 10.1 on pp. 45/8 3*

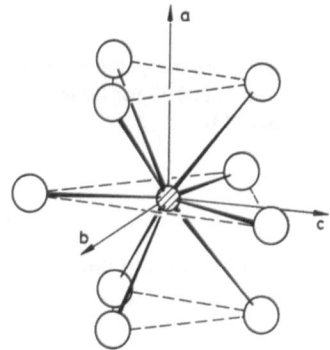

Fig. 17. Coordination polyhedron of ThS_2 [141].

The atomic positions are given for one set of four thorium atoms and two sets of four sulfur atoms by [141], see also [5, 60, 131]: $\pm(^{1}/_{4}, y, z)$; $\pm(^{1}/_{4}, ^{1}/_{2} + y, ^{1}/_{2} - z)$ with parameters of Th: $y = 0.24745 \pm 0.00003$, $z = -0.11982 \pm 0.00003$; S(1): $y = -0.1403 \pm 0.0002$, $z = -0.0698 \pm 0.0002$; S(2): $y = -0.0297 \pm 0.0002$, $z = 0.3338 \pm 0.0003$.

Each tetravalent thorium is coordinated to nine sulfur atoms (see **Fig. 17**) with an average distance of 2.95 Å [5], see also [2, 3]; the single values are Th-1 S(1) = 2.796 ± 0.002 Å, Th-2 S(2) = 2.988 ± 0.002 Å, Th-1 S(1) = 2.851 ± 0.002 Å, Th-1 S(2) = 2.948 ± 0.002 Å, Th-2 S(1) = 2.800 ± 0.002 Å, Th-2 S(2) = 3.235 ± 0.002 Å [141], see also [5, 132]. The smallest S-S distances are S(1)-S(1) = 3.26 Å, S(1)-S(2) = 3.27 Å, and 3.32 Å [5]. S-Th-S bond angles are tabulated in [141].

For an interpretation of crystal field effects performed on a Gd^{3+} doped single crystal of ThS_2, see [131, 132, 141].

10.1.5.3 Thermal Properties

Vaporization. Melting Point

ThS_2 loses sulfur above 1900°C [6]. Measured values for the vapor pressure are 43 Torr at 651°C, 163 Torr at 713°C, 382 Torr at 754°C [14].

The melting point of ThS_2 is 1905 ± 30°C [6], see also [2, 11, 17, 80, 122].

Heat Capacity and Thermodynamic Functions

The heat capacity of ThS_2 (78.54% Th, 21.43% S), corresponding to $ThS_{1.974}$ [13], measured in the temperature range 51 to 298 K, is shown in **Fig. 18** and Table 22 [137], see also [13, 17].

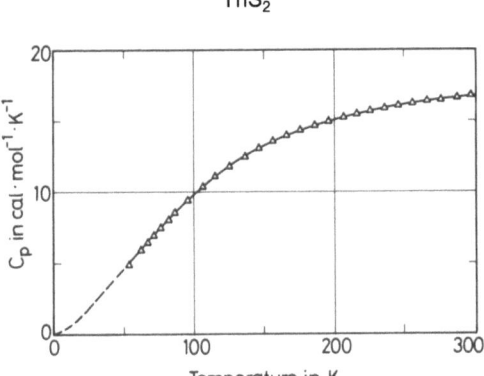

Fig. 18. Low-temperature heat capacity of ThS$_2$ [137].

Table 22
Low-Temperature Heat Capacity Data of ThS$_2$ (78.54% Th, 21.34% S) [137].

temperature in °C	heat capacity C_p in cal · mol^{-1} · K^{-1}	temperature in °C	heat capacity C_p in cal · mol^{-1} · K^{-1}
10	(0.16)	175	14.37
25	(1.73)	200	15.12
50	4.61	225	15.69
75	7.46	250	16.14
100	9.95	275	16.51
125	11.86	298.15	16.80
150	13.33		

The heat capacity at room temperature is C_p (at 298 K) = 16.80 ± 0.06 cal · mol^{-1} · K^{-1} [137], see also [13]. A further value is reported for the temperature range of 25 to 500°C to be C_p = 17.4 cal · mol^{-1} · K^{-1} [136], see also [13]. The following temperature dependences of the heat capacity are suggested: C_p (in cal · mol^{-1} · K^{-1}) = 16.02 + 2.63 × 10^{-3} T (298 to 2000 K) [52], see also [13], and C_p (in cal · mol^{-1} · K^{-1}) = 17.16 + 2.3 × 10^{-3} T (298 to 2188 K) [110], see also [13]. Further estimates are based on a value of the Debye temperature of Θ_D = 380 K [13] derived from the heat capacity measurements of [137], and an estimated value of a dilation contribution of 1.15 cal · mol^{-1} · K^{-1} at 1000 K [13]. The different results are summarized in **Fig. 19**, p. 38.

Values for entropy, S°, enthalpy, H$_T^\circ$ – H$_0^\circ$, and Gibbs free energy, –(G$_T^\circ$ – H$_0^\circ$)/T, based on measured heat capacity data from [137] are summarized in Table 23, p. 38, and values based on estimated heat capacity data of [13] are summarized in Table 24, p. 39.

Thermodynamic data of gaseous ThS$_2$(g) are only estimated and are quite tentative. The following values are estimated [13]: dissociation energy D° (S-ThS;0) ≈ 120 ± 10 kcal/mol,

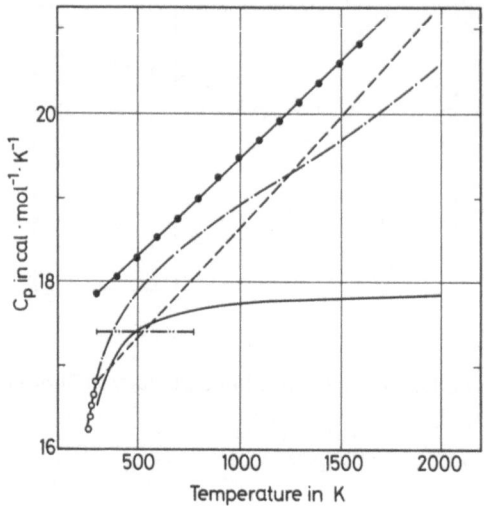

Fig. 19. Heat capacity of ThS$_2$(c) [13]. ○ results of [137], —··—··— results of [136] at 298 to 773 K,
—–—–— estimated by [52], —●●●— estimated by [110], ——— estimated C$_v$(l) (Θ_D = 380 K)
[13], —·—·— estimated C$_v$(l) + C(d) [13]. l = lattice, d = dilation.

Table 23
Thermodynamic Functions for Thorium Disulfide, ThS$_2$(c), in the Range 5 to 298 K [13].
Values in parentheses are less accurate than those in Table 22.

T in K	C$_p^\circ$ in cal · mol^{-1} · K^{-1}	S$^\circ$ in cal · mol^{-1} · K^{-1}	H$_T^\circ$ − H$_0^\circ$ in cal/mol	−(G$_T^\circ$ − H$_0^\circ$)/T in cal · mol^{-1} · K^{-1}
5	(0.018)	(0.0027)	(0.017)	(0.000)
10	(0.169)	(0.051)	(0.408)	(0.010)
15	(0.528)	(0.180)	(2.058)	(0.043)
20	(1.09)	(0.405)	(6.028)	(0.103)
25	(1.76)	(0.719)	(13.13)	(0.194)
30	(2.43)	(1.100)	(23.62)	(0.312)
40	(3.53)	(1.958)	(53.73)	(0.615)
50	4.64	2.863	94.49	0.973
60	5.78	3.810	146.63	1.366
80	7.99	5.780	284.7	2.222
100	9.93	7.777	464.3	3.134
150	13.33	12.512	1054.1	5.486
200	15.12	16.62	1769.9	7.770
250	16.15	20.11	2553.6	9.898
298.15	16.82	23.02	3348.1	11.787

corresponding to an atomization energy of D$^\circ$ (ThS$_2$, 298) ≈ 256 ± 15 kcal/mol, and enthalpy
of formation of ΔH_f°(ThS$_2$, g, 298) ≈ 18 ± 15 kcal/mol. More thermodynamic data of ThS$_2$(g) are
given in Table 25, p. 40.

Table 24
Thermodynamic Functions of ThS$_2$(c) in the Range 298 to 2000 K [13].
See note in Table 9 on p. 15.

T in K	C_p° in cal·mol^{-1}·K^{-1}	S° in cal·mol^{-1}·K^{-1}	$-(G_T^\circ - H_{298}^\circ)/T$ in cal·mol^{-1}·K^{-1}	$H_T^\circ - H_{298}^\circ$ in cal/mol	ΔH_f° in cal/mol	ΔG_f° in cal/mol	log K$_p$
298	16.800	23.000	23.000	0	−150000	−148485	108.841
300	16.840	23.104	23.000	31	−150001	−148476	108.163
400	17.480	28.042	23.669	1749	−151144	−147913	80.815
500	17.890	31.989	24.952	3519	−151941	−147014	64.259
600	18.160	35.275	26.406	5322	−152559	−145960	53.165
700	18.390	38.093	27.879	7150	−153058	−144827	45.216
800	18.560	40.559	29.313	8997	−179477	−146021	39.890
900	18.790	42.759	30.687	10865	−179255	−141851	34.446
1000	18.900	44.744	31.995	12749	−179045	−137707	30.095
1100	19.050	46.553	33.237	14647	−178849	−133583	26.540
1200	19.210	48.217	34.417	16560	−178664	−129476	23.581
1300	19.370	49.761	35.539	18489	−178491	−125385	21.079
1400	19.530	51.202	36.607	20434	−178327	−121305	18.936
1500	19.690	52.555	37.625	22395	−178174	−117237	17.081
1600	19.860	53.831	38.599	24372	−178031	−113179	15.459
1700	20.030	55.040	39.531	26367	−178711	−109096	14.025
1800	20.200	56.185	40.424	28370	−178534	−105007	12.749
1900	21.380	57.318	41.284	30465	−178299	−100927	11.609
2000	20.570	58.403	42.113	32579	−178080	−96863	10.585

References for 10.1 on pp. 45/8

Table 25
Thermodynamic Functions for Gaseous ThS$_2$(g) in the Range 298 to 3000 K [13].
See note in Table 9 on p. 15.

T in K	C$_p^\circ$ in cal·mol^{-1}·K^{-1}	S$^\circ$ in cal·mol^{-1}·K^{-1}	$-(G_T^\circ - H_{298}^\circ)/T$ in cal·mol^{-1}·K^{-1}	H$_T^\circ - H_{298}^\circ$ in cal/mol	ΔH_f° in cal/mol	ΔG_f° in cal/mol	log K$_p$
298	12.685	72.630	72.630	0	18000	4718	−3.458
300	12.698	72.709	72.630	23	17991	4635	−3.377
400	13.178	76.433	73.135	1319	16426	301	−0.164
500	13.426	79.402	74.101	2650	15190	− 3589	1.569
600	13.567	81.863	75.196	4000	14119	− 7234	2.635
700	13.655	83.961	76.302	5362	13154	−10723	3.348
800	13.714	85.789	77.376	6730	−13744	−16471	4.500
900	13.754	87.407	78.402	8104	−14016	−16795	4.078
1000	13.783	88.857	79.377	9481	−14313	−17088	3.735
1100	13.805	90.172	80.299	10860	−14636	−17351	3.447
1200	13.822	91.374	81.173	12242	−14982	−17583	3.202
1300	13.835	92.481	82.000	13624	−15356	−17785	2.990
1400	13.845	93.506	82.786	15008	−15753	−17956	2.803
1500	13.853	94.462	83.533	16393	−16176	−18098	2.637
1600	13.860	95.356	84.244	17779	−16624	−18212	2.488
1700	13.866	96.197	84.923	19165	−17913	−18263	2.348
1800	13.871	96.989	85.572	20552	−18352	−18272	2.218
1900	13.875	97.739	86.192	21940	−18824	−18254	2.100
2000	13.878	98.451	86.788	23327	−19332	−18212	1.990
2100	13.881	99.128	87.359	24715	−23277	−18015	1.875
2200	13.884	99.774	87.909	26103	−23961	−17749	1.763
2300	13.886	100.391	88.438	27492	−24648	−17449	1.658
2400	13.888	100.982	88.949	28881	−25341	−17123	1.559
2500	13.890	101.549	89.442	30269	−26037	−16765	1.466
2600	13.891	102.094	89.918	31659	−26735	−16382	1.377
2700	13.893	102.619	90.379	33048	−27436	−15970	1.293
2800	13.894	103.124	90.825	34437	−28141	−15531	1.212
2900	13.895	103.611	91.257	35827	−28849	−15068	1.136
3000	13.896	104.082	91.677	37216	−29560	−14581	1.062

10.1.5.4 Electrical Properties

ThS$_2$ is reported to be an insulator [6], see also [2, 81] or a semiconductor [98] with a room temperature resistivity of $1 \times 10^{10}\,\Omega \cdot$ cm (pressed powders at a pressure of 25000 psi) [98], also cited in [17], see also [101], or $5 \times 10^9\,\Omega \cdot$ cm [48].

ThS$_2$ was found to have a high Seebeck coefficient of the thermoelectric power of 700 µV/°C at 400°C [101], see also [17].

10.1.5.5 Magnetic Properties

ThS$_2$ has been reported to be diamagnetic [2], see also [17, 102] or not paramagnetic [6]. However, a weak temperature-independent paramagnetism was observed from susceptibility measurements, performed on sintered samples at 77 to 300 K, with a room temperature value of $\chi = 0.555 \times 10^{-6}$ cm^3/g [102].

10.1.5.6 Chemical Reactions

On Heating

ThS$_2$, brown to purple in color, was observed to lose sulfur above 1900°C [6], see also [121, 122]. ThS$_2$ is easily reduced to intermediate sulfides on heating in vacuum [6], see also [2]. For further details, see the chapter on "Thermal Properties", p. 36.

With Elements

ThS$_2$ is reported to be quite stable in air at room temperature [6], to react with oxygen at 400°C with slow formation of SO$_2$ [121], see also [122], or to be stable in a stream of oxygen up to 500°C [114]. The reaction was completed at 1300 to 1350°C with the formation of ThO$_2$ (see Table 26) [114]. No reaction was observed with ThS$_2$ and sulfur vapor at 400 to 800°C

Table 26
Oxidation of ThS$_2$ in an Oxygen Atmosphere (sulfur content of the original sulfide 21.0%) [114].

temperature in °C	S oxidized, in wt%, after 10 to 60 min						S content of the oxidized sulfide after 1 h in wt%
	10	20	30	40	50	60	
300	0	0	0	0	0	0	21.0
400	0	0	0	0	0	0	21.0
500	0	0	0	0	0	0	21.0
600	0.5	0.6	0.6	0.8	0.8	0.82	12.8
700	3.2	6.6	10.0	12.9	15.05	17.2	3.8
800	4.4	8.8	12.0	14.05	16.0	18.0	3.0
900	5.10	10.4	14.15	15.95	17.4	18.8	2.2
1000	5.8	11.2	15.2	16.5	18.0	19.05	1.05
1100	6.8	12.3	16.0	17.4	18.6	19.9	1.1
1200	7.9	15.4	17.5	18.5	19.5	20.2	0.8
1250	9.0	17.9	18.2	19.1	20.0	20.5	0.5
1300	10.0	18.0	18.9	19.9	20.4	21.0	0

[121], see also [122]. ThS$_2$ reacts with carbon and loses sulfur when heated up to 2800 °C on carbon in a nitrogen or hydrogen atmosphere. The reaction was observed to be complete after ½ h with the formation of ThC$_2$ [121], see also [122]. Otherwise, ThS$_2$, even if molten, is not attacked by carbon in an atmosphere of H$_2$S [2]. ThS$_2$ reacts with dry chlorine at 250 °C. No reaction was observed with bromine at room temperature, even not in the presence of dry air. Bromide and sulfate were observed at the reaction with bromine in the presence of water [121], see also [122].

No reaction was observed with ThS$_2$ and magnesium in an atmosphere of hydrogen at normal pressure at 800 °C [121], see also [122]. ThS$_2$ is reduced by reactive metals to ThS and the corresponding metal sulfides, as observed from melting experiments in sintered ThS$_2$ crucibles. Alkali and alkaline earth metals formed separate phases, beryllium a volatile sulfide, magnesium and barium attacked the crucible even at moderate temperatures (especially above 900 °C) while lanthanum, cerium, other rare earth metals, thorium, and uranium formed solid solutions [42], see also [2].

With Compounds

Well sintered ThS$_2$ is quite stable against water, even if boiling [6, 119]; reaction to ThO$_2$ and H$_2$S is reported at temperatures up to 200 °C [121], see also [122].

ThS$_2$ is only slowly dissolved in dilute acids at room temperature [6]. Dissolution readily occurs in >1 N HCl and HClO$_4$ [2]. A very slow attack occurred with cold HCl [121], see also [122]. Otherwise, there was no reaction observed with dilute as well as concentrated HCl even if hot. Rapid reaction was observed in dilute HNO$_3$, if warmed up, and in cold concentrated HNO$_3$ [1]. The formation of thorium nitrate, sulfur, and H$_2$S was observed in dilute (10%) HNO$_3$, but no formation of sulfate; dissolution in concentrated HNO$_3$ leads to the formation of thorium sulfate [121], see also [122]. Sulfate formation was also observed with aqua regia [121], see also [122]. ThS$_2$ was not attacked by dilute H$_2$SO$_4$, even if hot; only incomplete dissolution occurred in hot concentrated H$_2$SO$_4$ [1]. Otherwise, the formation of thorium sulfate and H$_2$S was observed to occur in the reaction with dilute H$_2$SO$_4$, and the formation of sulfate and SO$_2$ with hot concentrated H$_2$SO$_4$ [121], see also [122]. Dissolution in hydrofluoric acid leads to the formation of fluoride and H$_2$S [121], see also [122]. Only a slow reaction was observed with aqueous NaOH. Dissolution of ThS$_2$ in molten NaOH under air leads to the formation of ThO$_2$ and sulfur; limited formation of sulfate was observed [121], see also [122]. The values listed in Table 27 were obtained after dissolution in hot acids and NaOH. The observed reactions of ThS$_2$ with gaseous media are summarized in Table 28.

Table 27
Dissolution of ThS$_2$ in Hot Acids and Sodium Hydroxide after an Exposure of 30 min [119].

medium of dissolution	temperature in °C	content of sulfide in the residue in wt%
H$_2$SO$_4$ (1:1)	136	hydrolysis
HCl (D = 1.19)	112	0
H$_3$PO$_4$ (1:1)	100	1.8
H$_2$C$_2$O$_4$ (saturated)	100	20.1
CH$_3$COOH (glacial)	100	14.3
NaOH (20%)	100	7.38

Table 28
Reaction of ThS$_2$ with Gaseous Media (summarized from [121]).

gaseous medium	reaction observed
HCl, dry	reaction at $\geq 300\,°C$
H$_2$S	no reaction at 400 to 800 °C
SO$_2$	reaction at $\geq 400\,°C$ with formation of ThO$_2$ and sulfur; complete reaction at 500 °C within 1 h
NH$_3$	no reaction up to 1000 °C
CO$_2$	reaction at $\geq 400\,°C$; rapid reaction at $\geq 500\,°C$; ThS$_2$ + 2 CO$_2$ = ThO$_2$ + 2 S + 2 CO

H$_2$O$_2$ and KMnO$_4$ were observed to react with ThS$_2$ forming thorium sulfate. The reaction with K$_3$[Fe(CN)$_6$] led to the formation of ThO$_2$ and sulfur [121], see also [122].

Sintered ThS$_2$ crucibles are reported to be useful containers for reactive halides up to 900 °C [2] or 1000 °C [42], see also [44].

With Th$_7$S$_{12}$, ThS$_2$ forms a eutectic composition which melts at 1765 $\pm$ 25 °C [2], or 1800 to 1850 °C [11].

For further information, especially from the older literature, see "Thorium" 1955, pp. 281/2.

10.1.6 Dithorium Pentasulfide, Th$_2$S$_5$

10.1.6.1 Formation and Preparation

Th$_2$S$_5$ was prepared by the reaction of thorium powder, obtained from decomposition of thorium hydride, and sulfur in excess. The reaction was carried out in a sealed silica tube at 400 to 420 °C for 2 to 3 weeks to ensure complete reaction and reasonable growth of crystals. Th$_2$S$_5$, brick-red with slight purple tint in color, then was obtained after sublimation of the excess sulfur at 250 to 300 °C [18], see also [12, 17]. One attempt to prepare a polysulfide led to the formation of ThS$_{2.36}$ or Th$_3$S$_7$ [1], see also "Thorium" 1955, p. 282, but a Th$_3$S$_7$ phase has not been confirmed.

Small single crystals of Th$_2$S$_5$ were obtained by a gas-phase transport method using bromine as the transporting agent within the temperature gradient of 850 to 750 °C [139].

The enthalpy of formation of Th$_2$S$_5$ is ΔH_f°(Th$_2$S$_5$, c, 298) $= -152 \pm 10$ kcal/mol, taking ΔH_f° (ThS$_2$, c, 298) $= -150$ kcal/mol into account [13].

10.1.6.2 Crystallographic Properties

Th$_2$S$_5$ is orthorhombic (pseudo-tetragonal) with four molecules per unit cell; the space group is Pcnb-D$_{2h}^{14}$ (No. 60) [139], see also [13, 21]. This space group was determined from single crystal measurements using the Weissenberg method. A tetragonal symmetry, space group P4$_2$/n-C$_{4h}^4$ (No. 86) or P4$_2$nmc-D$_{4h}^{15}$ (No. 137), was derived from powder diffraction pattern, leading to the lattice parameters a = 5.43 Å, c = 10.15 Å [18], see also [12, 13, 17, 67], or a = 10.80 $\pm$ 0.01 Å, c = 10.20 $\pm$ 0.01 Å [12], see also [13, 123] or a = 5.406 Å, c = 10.13 Å [13]. A X-ray density of 6.90 g/cm^3 was calculated from this tetragonal data, assuming the Th$_2$S$_5$ to contain two molecules per unit cell. The pycnometric density was 6.92 g/cm^3 [18].

References for 10.1 on pp. 45/8

The lattice parameters of the orthorhombic unit cell are a = 7.623 Å, b = 7.677 Å, c = 10.141 Å [139], see also [13, 21]. From this the X-ray density was calculated to be 6.99 g/cm³ [139], see also [21].

Assuming the space group Pcnb, the thorium atoms are in the general positions of the 8(d) sites, and the sulfur atoms are situated in two positions of the 8(d) sites, S(1) and S(2), and one position of the 4(c) sites, S(3), with the following parameters:

Th:	x = 0.23171 (9)	y = 0.02183 (9)	z = 0.14600 (8)
S(1):	x = 0.3615 (7)	y = 0.3876 (7)	z = 0.6077 (6)
S(2):	x = 0.9028 (7)	y = 0.8484 (7)	z = 0.5996 (6)
S(3):	x = 0	y = 0.25	z = 0.7528 (9)

The interatomic distances are (in Å)

Th-Th = 4.369 (1)		S(1)-S(1) = 2.988 (7)
Th-Th = 4.389 (1)		S(1)-S(1) = 3.494 (7)
Th-S(1) = 2.861 (4)		S(2)-S(2) = 2.117 (7)
Th-S(1) = 2.877 (4)		S(2)-S(2) = 3.420 (7)
Th-S(1) = 2.922 (4)		S(2)-S(3) = 3.498 (4)
Th-S(1) = 2.933 (4)		S(2)-S(3) = 3.531 (4)
Th-S(2) = 2.949 (4)		S(1)-S(2) = 2.858 (5)
Th-S(2) = 2.983 (4)		S(1)-S(2) = 3.430 (5)
Th-S(2) = 3.123 (4)		S(1)-S(2) = 3.484 (5)
Th-S(2) = 3.163 (4)		S(1)-S(2) = 3.492 (5)
Th-S(3) = 2.902 (2)		S(1)-S(3) = 3.295 (5)
Th-S(3) = 2.920 (2)		S(1)-S(3) = 3.298 (4)

The short distance S(2)-S(2) = 2.117 Å confirms the Th_2S_5 to be a polysulfide [21]. The coordination polyhedron is shown in **Fig. 20**.

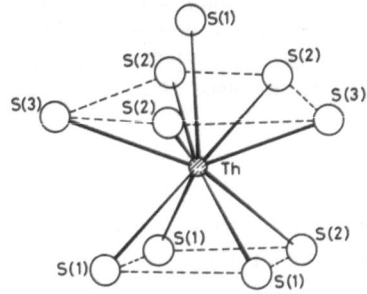

Fig. 20. Coordination polyhedron of Th_2S_5 [21].

10.1.6.3 Thermal Properties

Th_2S_5 is reported to dissociate at about 600°C [12].

An estimated value for the room temperature entropy of Th_2S_5 is S° (Th_2S_5, c, 298) = 27 ± 3 cal · mol⁻¹ · K⁻¹ [13].

10.1.6.4 Electrical Properties

Th$_2$S$_5$ is reported to be a semiconductor with a room temperature resistivity of $5 \times 10^9 \, \Omega \cdot$ cm (powders pressed at 25 000 psi) [98], see also [17], or $2 \times 10^9 \, \Omega \cdot$ cm [48].

10.1.6.5 Chemical Reactions

Th$_2$S$_5$, red in color, is reported to dissociate at about 600 °C [12]. Th$_2$S$_5$ was degraded in vacuum at 950 °C to form ThS$_2$ [105].

In an atmosphere of hydrogen, Th$_2$S$_5$ is reduced to ThS$_2$ at 950 °C within 3 h. Th$_2$S$_5$ readily reacts with chlorine at temperatures below 300 °C. Th$_2$S$_5$, like all thorium sulfides, is reported to be reduced by alkali metals to the metal [105].

Th$_2$S$_5$ is stable against water, even if boiling. No reaction was observed with concentrated HCl and 2 N H$_2$SO$_4$, even if boiling. Th$_2$S$_5$ reacts with cold concentrated HNO$_3$ and with boiling H$_2$O$_2$ (50 vol%). No reaction was found with a solution of 5 N NaOH [105].

References for 10.1:

[1] E. F. Strotzer, M. Zumbusch (Z. Anorg. Allgem. Chem. **247** [1941] 415/28). — [2] L. Brewer, Le Roy A. Bromley, P. Gilles, N. Lofgren (AECD-2242 [1948] 1/47; N.S.A. **1** [1948] No. 1098). — [3] W. H. Zachariasen (Acta Cryst. **2** [1949] 288/91). — [4] A. S. Khan, D. T. Peterson (J. Less-Common Metals **50** [1976] 103/6). — [5] W. H. Zachariasen (Acta Cryst. **2** [1949] 291/6).

[6] E. D. Eastman, L. Brewer, Le Roy A. Bromley, P. W. Gilles, N. L. Lofgren (J. Am. Chem. Soc. **72** [1950] 4019/23). — [7] B. Griveau, J. P. Marcon, J. P. Gatesoupe (J. Inorg. Nucl. Chem. **34** [1972] 1535/9). — [8] H. Adachi, Y. Henmi, S. Imoto (Technol. Rept. Osaka Univ. **22** [1972] 121/5; C.A. **77** [1972] No. 144710). — [9] S. Aronson (J. Inorg. Nucl. Chem. **29** [1967] 1611/7). — [10] P. D. Shalek, J. H. Handwerk (ANL-FGF-397 [1962] 1/18; C.A. **60** [1964] 15546).

[11] L. Brewer, Le Roy A. Bromley, P. Gilles, N. Lofgren (CC-2664 [1945] 1/8; N.S.A. **10** [1956] No. 5257). — [12] J. P. Marcon (CEA-R-3919 [1969] 1/99; C.A. **72** [1970] No. 106632). — [13] F. Grønvold, J. Drowart, E. F. Westrum Jr. (in: F.L. Oetting, The Chemical Thermodynamics of Actinide Elements and Compounds, IAEA, Vienna 1984, pp. 13/27). — [14] G. V. Samsonov, N. M. Popova (Zh. Obshch. Khim. **27** [1957] 3/10; J. Gen. Chem. [USSR] **27** [1957] 1/7; C.A. **1957** 13629, **1958** 14404). — [15] L. Brewer, N. L. Lofgren, E. D. Eastman (U.S. Appl. 791468 [1950]; Offic. Gaz. U.S. Patent Office **641** [1950] 1347; C.A. **1952** 2258).

[16] L. Brewer, N. L. Lofgren, E. D. Eastman (U.S. Appl. 791469 [1950]; Offic. Gaz. U.S. Patent Office **641** [1950] 1347; C.A. **1952** 2259). — [17] R. M. Dell, M. Allbutt (AERE-R-4253 [1963] 1/48; C.A. **59** [1963] 204). — [18] J. Graham, F. K. McTaggart (Australian J. Chem. **13** [1960] 67/73). — [19] A. R. Moodenbaugh, D. C. Johnston, R. Viswanathan, R. N. Shelton, L. E. de Long, W. A. Fertig (J. Low Temp. Phys. **33** [1978] 175/203). — [20] A. R. Moodenbaugh (Diss. Univ. California 1975, pp. 1/169; N.S.A. **33** [1976] No. 30308).

[21] H. Noël, M. Potel (Acta Cryst. B **38** [1982] 2444/5). — [22] P. D. Shalek (J. Am. Ceram. Soc. **46** [1963] 155/61). — [23] J. H. Handwerk, O. L. Kruger (Nucl. Eng. Design **45** [1971] 397/408). — [24] P. D. Shalek (ANL-FGF-247 [1961] 1/14; C.A. **61** [1964] 12899). — [25] H. E. Flotow, D. W. Osborne, R. R. Walters (J. Chem. Phys. **55** [1971] 880/6).

[26] P. D. Shalek, G. D. White, J. H. Handwerk (ANL-FGF-361 [1962] 1/17; C.A. **59** [1963] 12361). – [27] P. D. Shalek, J. H. Handwerk (Spec. Ceram. No. 2 [1962/63] 1/10). – [28] J. Barghusen (React. Fuel Process. **8** No. 1 [1964/65] 50/64). – [29] B. A. Rogers (ISC-82 [1950] 1/30; N.S.A. **10** [1956] No. 4305). – [30] M. Haessler, C. H. de Novion (J. Phys. C **10** [1977] 589/602).

[31] S. Aronson, E. Cisney, K. A. Gingerich (J. Am. Ceram. Soc. **50** [1967] 248/52). – [32] J. Flahaut (Silicon Sulfur Phosphates Colloq., Münster, Ger., 1954 [1955], pp. 165/7; C.A. **1958** 3580). – [33] G. H. B. Lovell, D. R. Perels, E. J. Britz (J. Nucl. Mater. **39** [1971] 303/10). – [34] R. Didchenko, L. M. Litz (J. Electrochem. Soc. **109** [1962] 247/50). – [35] R. Didchenko, L. M. Litz (U.S. 3086925 [1963]; C.A. **58** [1963] 13585).

[36] P. D. Shalek (Trans. Am. Nucl. Soc. **5** [1962] 244/5). – [37] P. D. Shalek (ANL-6677 [1962] 107/9; N.S.A. **17** [1963] No. 41333). – [38] Y. Baskin (ANL-6856 [1964] 27/30; N.S.A. **18** [1964] No. 37115). – [39] R. B. Holden (Ceramic Fuel Elements, Gordon & Breach, New York 1966, pp. 167/83). – [40] P. D. Shalek (ANL-FGF-325 [1962] 1/11; N.S.A. **16** [1962] No. 32090).

[41] P. D. Shalek (ANL-FGF-347 [1962] 1/24; N.S.A. **17** [1963] No. 18739). – [42] E. D. Eastman, L. Brewer, Le Roy A. Bromley, P. W. Gilles, N. L. Lofgren (J. Am. Ceram. Soc. **34** [1951] 128/34). – [43] G. Bieler (Silicates Ind. **18** [1953] 234/6; C.A. **1953** 12781). – [44] L. Brewer, Le Roy A. Bromley, P. W. Gilles, N. Lofgren (AECD-2253 [1948] 1/32; N.S.A. **1** [1948] No. 1099). – [45] C. d'A. Hunt, H. R. Smith Jr. (U.S. 3 199 947 [1965]; C.A. **63** [1965] 14521).

[46] J. Fuger (MTP [Med. Tech. Publ. Co.] Intern. Rev. Sci. Inorg. Chem. Ser. One **7** [1972] 157/210). – [47] F. H. Norton (AECD-2237 [1948] 1/5; N.S.A. **1** [1948] No. 1097). – [48] G. V. Samsonov (Poroshkovaya Met. **1962** No. 4, pp. 11/9; Soviet Powder Met. Metal Ceram. **1962** No. 4, pp. 237/43; N.S.A. **17** [1963] No. 8080). – [49] S. Aronson, A. Auskern (BNL-50023 [1966] 146/8; N.S.A. **21** [1967] No. 35701). – [50] Le Roy Eyring, E. F. Westrum Jr. (J. Am. Chem. Soc. **75** [1953] 4802/3).

[51] E. D. Cater, R. J. Thorn, R. R. Walters (Met. Soc. AIME Inst. Metals Div. Spec. Rept. Ser. **10** No. 13 [1964] 237/44). – [52] K. C. Mills (Thermodynamic Data for Inorganic Sulfides, Selenides and Tellurides, Butterworth, London 1974). – [53] M. Ader (ANL-7921 [1972] 17/8; N.S.A. **26** [1972] No. 47738). – [54] P. A. G. O'Hare, M. Ader, W. N. Hubbard, G. K. Johnson, J. L. Settle (Thermodyn. Nucl. Mater. Proc. 4th Symp., Vienna 1974 [1975], Vol. 2, pp. 439/53). – [55] M. Ader (ANL-7876 [1972] 21/6; C.A. **77** [1972] No. 119023).

[56] M. Ader (ANL-7821 [1971] 16/7; N.S.A. **25** [1971] No. 44367). – [57] H. Rand (At. Energy Rev. Spec. Issue No. 5 [1975] 7/85). – [58] D. D. Wagman, R. H. Schumm, V. B. Parker (NBSIR-77-1300 [1977] 1/94; PB-273171 [1977] 1/94; INIS Atomindex **9** [1978] No. 410598). – [59] K. N. Strafford, G. R. Winstanley, J. M. Harrison (Werkstoffe Korrosion **25** [1974] 487/96). – [60] D. J. Lam, J. B. Darby Jr., M. V. Nevitt (in: A. J. Freeman, J. B. Darby Jr., The Actinides: Electronic Structure and Related Properties, Vol. 2, Academic, New York 1974, pp. 119/84).

[61] M. Allbutt, R. M. Dell (J. Inorg. Nucl. Chem. **30** [1968] 705/10). – [62] M. Allbutt, A. R. Junkison, R. M. Dell (Met. Soc. AIME Inst. Metals Div. Spec. Rept. Ser. **10** No. 13 [1964] 65/81). – [63] U. Benedict, J. C. Spirlet, L. Gerward, J. Staun Olsen (J. Less-Common Metals **98** [1984] 301/7). – [64] L. I. Katzin (Natl. Nucl. Energy Ser. Div. IV B **14** [1954] 62/102). – [65] H. Matzke, R. Lindner (Atomkernenergie **9** [1964] 2/46).

[66] I. Krivy (UJV-1738 [1967] 1/19; C.A. **68** [1968] No. 35069). – [67] R. M. Dell, N. J. Bridger (MTP [Med. Tech. Publ. Co.] Intern. Rev. Sci. Inorg. Chem. Ser. One **7** [1972] 211/74). – [68] M. Allbutt, R. M. Dell (J. Nucl. Mater. **24** [1967] 1/20). – [69] R. J. Beals (Argonne Natl. Lab. Rev. **2** No. 1 [1965] 14/5; N.S.A. **19** [1965] No. 13288). – [70] M. Tetenbaum (J. Appl. Phys. **35** [1964] 2468/72).

[71] S. Aronson, A. Ingraham (J. Nucl. Mater. **24** [1967] 74/9). – [72] F. A. Wedgwood (J. Phys. C **7** [1974] 3203/18). – [73] J. Danan, C. H. de Novion, Y. Guerin, F. A. Wedgwood, M. Kuznietz (J. Phys. [Paris] **37** [1976] 1169/86). – [74] K. G. Rajan, R. Krishnan, A. Sequeira,

G. Venkataraman (Proc. Nucl. Phys. Solid State Phys. Symp. C **16** [1973] 160). — [75] K. P. Thakur (Australian J. Phys. **30** [1977] 325/34).

[76] P. S. Bakhshi, V. K. Jain, J. Shanker (J. Inorg. Nucl. Chem. **43** [1981] 901/9). — [77] K. P. Thakur (J. Phys. Chem. Solids **41** [1980] 465/72). — [78] S. Aronson, A. Ingraham (BNL-50023 [1966] 1/191, 148; N.S.A. **21** [1967] No. 35701). — [79] T. E. Hanley (J. Appl. Phys. **21** [1950] 1193). — [80] T. A. Badaeva (in: E. M. Potapova, Stroenie Splavov Nekotorykh System s Uranom i Toriem, Gosatomizdat., Moscow [1961], pp. 339/57 [The Structure of Alloys of Certain Systems Containing Uranium and Thorium]; AEC-TR-5834 [1963] 321/36; N.S.A. **16** [1962] No. 30877).

[81] E. Lubatti, S. Pappalardo, U. Mirarchi (Termotechnica [Milan] **9** [1966] 540/51). — [82] M. S. Farkas, A. A. Bauer, R. F. Dickerson (BMI-1568 [1962] 1/20; C.A. **56** [1962] 13732). — [83] R. F. Voitovich, N. I. Shakhanoya (Poroshkovaya Met. **1967** No. 3, pp. 75/9; Soviet Powder Met. Metal Ceram. **1967** No. 3, pp. 225/8; C.A. **67** [1967] No. 26516). — [84] J. Danan, B. Griveau, J. P. Marcon, J. P. Gatesoupe, C. H. de Novion (Conf. Dig. Inst. Phys. [London] No. 3 [1971] 176/9). — [85] J. Danan (CEA-R-4453 [1973] 1/138; C.A. **80** [1974] No. 20177).

[86] E. F. Westrum Jr., F. Grønvold (Thermodyn. Nucl. Mater. Proc. Symp., Vienna 1962 [1963], pp. 3/36, 23/36; SM-26-30 [1962/63] 1/13; C.A. **62** [1965] 15493). — [87] D. L. McElroy (ORNL-3670 [1964] 23/6; TID-4500-34th Ed. [1964]; N.S.A. **18** [1964] No. 44148). — [88] R. W. Powell (J. Sci. Instr. **34** [1957] 485/92). — [89] T. G. Kollie, D. L. McElroy, R. S. Graves, W. Fulkerson (ORNL-P-148 [1964] 1/47; N.S.A. **18** [1964] No. 35990). — [90] Anonymous (React. Mater. **7** [1964/65] 211/29).

[91] S. Imoto, H. Adachi, T. Hori (J. Nucl. Sci. Technol. [Tokyo] **12** [1975] 711/6; C.A. **84** [1976] No. 79897). — [92] F. A. Wedgwood, C. de Novion (4th Actinides Colloq., Harwell, Engl., 1974; AERE-R-7961 [1975] 66/73; N.S.A. **32** [1975] No. 20197). — [93] W. P. Ellis (in: A. J. Freeman, J. B. Darby Jr., The Actinides; Electronic Structure and Related Properties, Vol. 2, Academic, New York 1974, pp. 345/67). — [94] M. Tetenbaum (ANL-6856 [1964] 48/52; N.S.A. **18** [1964] No. 37115). — [95] Argonne National Laboratories (ANL-6875 [1964] 1/192; N.S.A. **18** [1964] No. 33678).

[96] M. B. Brodsky, A. J. Arko, A. R. Harvey (in: A. J. Freeman, J. B. Darby Jr., The Actinides: Electronic Structure and Related Properties, Vol. 2, Academic, New York 1974, pp. 185/264). — [97] M. Tetenbaum, F. Mrazek (ANL-6800 [1963] 420/2; N.S.A. **18** [1964] No. 44493). — [98] F. K. McTaggart (Australian J. Chem. **11** [1958] 471/80). — [99] Argonne National Laboratories (ANL-6766 [1963] 128/30; TID-4500-22nd Ed. [1963]; N.S.A. **17** [1963] No. 40421). — [100] C. H. De Novion, M. Konczykowski, M. Haessler (J. Phys. C **15** [1982] 1251/60).

[101] C. B. Jordan (EOS-1592-FINAL [1962] 1/43; N.S.A. **16** [1962] No. 10524). — [102] H. Adachi, S. Imoto (Technol. Rept. Osaka Univ. **23** [1973] 425/9; C.A. **81** [1974] No. 43068). — [103] G. Jaeger (Metall **9** [1955] 358/66). — [104] S. Fried, E. F. Westrum Jr., H. L. Baumbach, P. L. Kirk (J. Inorg. Nucl. Chem. **5** [1958] 182/9). — [105] J. Baer, F. K. McTaggart (Australian J. Chem. **11** [1958] 458/70).

[106] E. D. Eastman, L. Brewer, Le Roy A. Bromley, P. W. Gilles, N. L. Lofgren (M-4409 [1945] 1/7; C.A. **1961** 17422). — [107] Le Roy Eyring, E. F. Westrum Jr. (AECD-3519 [1953] 1/7; UCRL-2175 [1953] 1/7; N.S.A. **7** [1953] No. 3347). — [108] S. Fried, W. H. Zachariasen (Proc. 1st Intern. Conf. Peaceful Uses At. Energy, Geneva 1955 [1956], Vol. 7, pp. 235/44). — [109] A. I. Moskvin (Radiokhimiya **15** [1973] 353/62; Soviet Radiochem. **15** [1973] 356/63; C.A. **79** [1973] No. 97695). — [110] I. Barin, O. Knacke, O. Kubaschewski (Thermochemical Properties of Inorganic Substances, Suppl., Springer, Berlin 1977, pp. 1/861).

[111] S. Toesca, F. Le Boete, J.-C. Colson, D. Delafosse (Compt. Rend. C **268** [1969] 1099/102). — [112] G. V. Samsonov, G. N. Dubrovskaya (Zh. Prikl. Khim. **36** [1963] 1615/8; J. Appl. Chem. [USSR] **36** [1963] 1555/7; C.A. **60** [1964] 8922). — [113] G. V. Samsonov, G. N. Dubrovskaya (At. Energ. [USSR] **15** [1963] 428/30; Soviet At. Energy **15** [1964] 1191/3; N.S.A.

18 [1964] No. 8342). — [114] G. N. Dubrovskaya, I. N. Godovannaya (Zh. Analit. Khim. **19** [1964] 993/6; J. Anal. Chem. [USSR] **19** [1964] 922/4; C.A. **61** [1964] 12635). — [115] L. Brewer, N. L. Lofgren, E. D. Eastman (U.S. Appl. 791467 [1950]; Offic. Gaz. U.S. Patent Office **641** [1950] 1346; C.A. **1952** 2258).

[116] O. Kubaschewski, E. L. Evans, C. B. Alcock (Metallurgical Thermochemistry, Pergamon, Oxford 1967). — [117] W. H. Zachariasen (AECD-2141 [1946] 1/24; ANL-FWHZ-161 [1946] 1/24; N.S.A. **1** [1948] No. 723). — [118] N. M. Griesenauer, M. S. Farkas, F. A. Rough (BMI-1680 [1964] 1/32; C.A. **62** [1965] 2454). — [119] G. N. Dubrovskaya (Zh. Neorgan. Khim. **16** [1971] 12/5; Russ. J. Inorg. Chem. **16** [1971] 6/8; C.A. **74** [1971] No. 68454). — [120] J. J. Berzelius (Physik Chem. [2] **16** [1829] 385/415).

[121] M. Picon (Compt. Rend. **195** [1932] 957/9). — [122] M. Picon (Bull. Soc. Chim. France [4] **53** [1933] 166/70). — [123] J. P. Marcon (Compt. Rend. C **265** [1967] 235). — [124] H. Lipkind, A. S. Newton (TID-5223-Pt. 1 [1952] 398/404; C.A. **1957** 16167). — [125] A. S. Newton, O. Johnson (U.S. Appl. 787850 [1951]; Offic. Gaz. U.S. Patent Office **651** [1951] 615/6; C.A. **1952** 7723).

[126] L. Brewer, Le Roy A. Bromley, E. D. Eastman (U.S. Appl. 791466 [1950]; Offic. Gaz. U.S. Patent Office **641** [1950] 1346; C.A. **1952** 2258). —[127] A. K. Srivastava, R. K. Agarwal (Thermochim. Acta **68** [1983] 121/3). — [128] T. Ishimori, K. Ueno, E. Akatsu (J. Nucl. Sci. Technol. [Tokyo] **10** [1973] 95/100; C.A. **78** [1973] No. 126371). — [129] T. Ishimori, K. Ueno, E. Wagakatsu, M. Hoshi, M. Kawasaki (Japan. 70-33524 [1970] from C.A. **74** [1971] No. 55805). — [130] G. Amoretti, D. C. Giori, V. Varacca, J. C. Spirlet, J. Rebizant (Phys. Rev. [3] B **20** [1979] 3573/7).

[131] G. Amoretti, D. C. Giori, V. Varacca (Z. Naturforsch. **36a** [1981] 1163/8). — [132] G. Amoretti, C. Fava, V. Varacca (Z. Naturforsch. **37a** [1982] 536/45). — [133] D. E. Wilcox, Le Roy A. Bromley (Ind. Eng. Chem. **55** No. 7 [1963] 32/9). — [134] W. Brelvi (UCRL-16865 [1966] 1/92; N.S.A. **20** [1966] No. 35482). — [135] H. Bloom (Diss. Univ. London 1957).

[136] H. Hartmann, D. Mootz, Th. Nentwich (Angew. Chem. **73** [1961] 172/3). — [137] E. G. King, W. W. Weller (U.S. Bur. Mines Rept. Invest. No. 5485 [1959] 1/5; C.A. **1959** 15743). — [138] C.B. Jordan (AD-277002 [1964] 1/39; C.A. **61** [1964] 1364). — [139] H. Nöel (J. Inorg. Nucl. Chem. **42** [1980] 1715/7). — [140] U. Benedict, J. C. Spirlet, L. Gerward, J. S. Olsen (DESY-SR-83-22 [1983] 1/16; C.A. **101** [1984] No. 201919).

[141] G. Amoretti, G. Calestani, D. C. Giori (Z. Naturforsch. **39a** [1984] 778/82).

10.2   Compounds of Thorium with Sulfur and Oxygen

In this chapter only the one existing thorium oxide sulfide, ThOS, is dealt with. Thorium sulfites and sulfates are described in Chapters 10.5 and 10.6, pp. 57 and 63, respectively.

10.2.1   Thorium Oxide Sulfide, ThOS

10.2.1.1   Formation and Preparation

ThOS was prepared for the first time in 1894 [1]. Special information for the older preparative techniques is given in "Thorium" 1955, pp. 282/3.

Reaction of Thorium Oxide with H₂S

The reaction of thorium oxide with H_2S, which gives ThOS, water, and gaseous sulfur, was carried out in the absence of carbon in a refractory oxide tube at 1300°C [2], at 1100°C [3], and at 900 to 1100°C in a porcelain boat placed in a silica tube [4], see also [5]. Impurities of free sulfur found in the reaction product were removed by dissolution in dilute hydrochloric acid [3]. The reaction products obtained at different reaction temperatures are given in Table 29. The same reaction carried out in the presence of carbon — in a carbon apparatus at 1300°C [2], or in carbon crucibles at about 1500°C [6], or at 800 to 1000°C [4], see also [5, 7] — leads to the formation of carbon disulfide, CS_2, as an intermediate, which promotes the reaction. The reaction products obtained at different reaction temperatures are given in Table 20, p. 33.

Table 29
Chemical and Phase Composition of the Products of the Reaction between ThO_2 and H_2S in Absence of Carbon (time of reaction 1 h) [4].

temperature in °C	chemical composition in %					phase composition
	total Th	total S	free S	total Th + total S	oxygen (by difference)	
500	88	0	0	88	12	ThO_2 + ThOS
600	87.5	7.2	0.2	94.7	5.3	
700	86.6	7.8	0.2	94.4	5.6	
800	85.1	9.3	0.2	94.4	5.6	
900	84.7	9.9	0.1	94.6	5.4	ThOS
1000	83.2	11.5	0.1	94.7	5.3	
1100	82.6	12.5	0.1	95.1	4.9	
1200	81.5	17.7	0.1	99.2	0.8	$ThS_{1.7}$
1300	81.3	17.9	0.1	99.2	0.8	

Reaction of Thorium Oxide with Sulfur

The reaction of thorium oxide with sulfur vapor (vaporized at 600°C) was carried out at 1100°C in a special apparatus as shown in **Fig. 21**, p. 50. The sulfur was carried into the reaction zone in a stream of argon. Impurities of other thorium sulfides were removed by dissolution in dilute hydrochloric acid. The ThOS obtained was yellow in color [8].

Reaction of Thorium Oxide with ThS₂

The reaction of thorium oxide with ThS_2, placed in an aluminium crucible and sealed in an evacuated silica tube at 1000°C for three days, leads to almost pure ThOS with a total amount of impurities of 1.3 at.% per molecule of the compound [9].

Single Crystals

Single crystals of ThOS were grown from ThOS powder using a transport method with iodine as the transporting agent (2 atm, at 900°C), sealed in an evacuated (10^{-6} Torr) quartz

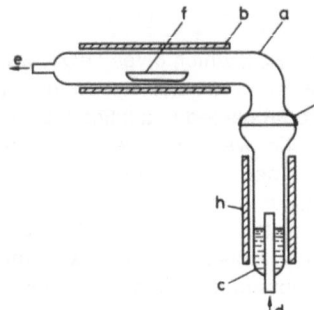

Fig. 21. Schematic apparatus for the preparation of ThOS [8]. a = reaction tube, b = heating element, c = vaporization of sulfur, d = argon inlet, e = exit, f = reaction crucible, g = spherical coupling, h = heating element (600°C).

tube. Crystals of 2 mm in length were grown after two weeks at temperatures of 900°C (hot end) to 850°C (cool end) [10]. ThOS crystals were also obtained from the elements (Th, S), mixed with a third element (Si, Ge, P, As, Sb, or Bi), by a transport method with bromine or iodine as the transporting agent (3 to 6 mg/cm³). The reactants are placed in uncoated quartz ampules at temperatures of 900 to 1050°C (hot end), the other side 50 to 100°C cooler; reaction occurs within one week. With this method the SiO_2 is reduced to elemental silicon (which may react with the third element to form SiP, SiAs, etc.) and the oxygen is used for the formation of ThOS. The crystals were transparent, orange-yellow in color, up to 3 mm in size, and most had a pseudo-hexagonal habit [11], also cited in [12]. The ThOS crystals were also obtained in the absence of the third element if the starting material corresponded to the composition ThS_2, provided the silica tube was uncoated [11].

Enthalpy of Formation

The enthalpy of formation was measured to be $\Delta H_f = -206 \pm 6$ kcal/mol (at 25°C) [6], see also [13]. An estimated value is $\Delta H_f = -210 \pm 20$ kcal/mol (at 1500 K) [2].

10.2.1.2 Crystallographic Properties

ThOS is tetragonal (PbFCl type) with two molecules per unit cell; the space group is P4/nmm-D_{4h}^7 (No. 129) [14], see also [10,12]. The measured lattice parameters are summarized in Table 30.

Table 30
Measured Lattice Parameters of ThOS.

a	c	Ref.
3.955 ± 2 kX	6.733 ± 4 kX	[14]; also cited in [11, 15 to 17]
3.963 ± 2 Å	6.747 Å	[2] from [14]; also cited in [12]
3.90 Å	6.72 Å	[8]
3.973 Å	6.733 Å	[10] (single crystal)

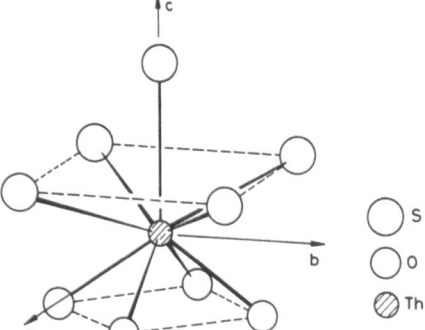

Fig. 22. Coordination polyhedron of ThOS [9].

The X-ray density was calculated to be 8.78 g/cm^3 [14], see also [2]. The atomic positions are 2 Th in ($^1/_2$, 0, u_1); (0, $^1/_2$, $\bar{u}_1$); 2 O in (0, 0, 0); ($^1/_2$, $^1/_2$, 0); 2 S in ($^1/_2$, 0, u_2); (0, $^1/_2$, $\bar{u}_2$) with u_1 = 0.200 ± 6, u_2 = 0.647. The interatomic distances are Th-4 O = 2.40 Å, Th-5 S = 3.00 Å [14]. The coordination polyhedron is shown in **Fig. 22**.

10.2.1.3 Heat Capacity and Entropy

The heat capacity, C_p, of ThOS was measured within the temperature range of 5 to 300 K. The measured values of C_p and calculated values for the entropy, S, are summarized in Table 31. The room temperature values are C_p = 67.25 J · mol^{-1} · K^{-1} and S = 76.34 J · mol^{-1} · K^{-1}, both at 298.15 K [9].

Table 31
Thermodynamic Functions for ThOS [9].

T in K	C_p in J · mol^{-1} · K^{-1}	S in J · mol^{-1} · K^{-1}	T in K	C_p in J · mol^{-1} · K^{-1}	S in J · mol^{-1} · K^{-1}
5.0	0.00	0.00	140.0	44.60	33.39
10.0	0.13	0.03	150.0	47.03	36.55
15.0	0.66	0.17	160.0	49.25	39.66
20.0	1.59	0.48	170.0	51.29	42.71
25.0	2.86	0.96	180.0	53.17	45.69
30.0	4.41	1.62	190.0	54.92	48.62
35.0	6.17	2.43	200.0	56.55	51.47
40.0	8.09	3.38	210.0	58.08	54.27
45.0	10.09	4.44	220.0	59.50	57.01
50.0	12.15	5.61	230.0	60.82	59.68
60.0	16.37	8.20	240.0	62.01	62.29
70.0	20.61	11.04	250.0	63.10	64.85
80.0	24.76	14.07	260.0	64.09	67.34
90.0	28.70	17.21	270.0	65.01	69.78
100.0	32.41	20.43	280.0	65.85	71.92
110.0	35.86	23.68	290.0	66.64	74.24
120.0	39.04	26.94	300.0	67.37	76.52
130.0	41.95	30.18			

The total heat capacity of ThOS consists of a dilation term, C_{dil}, a lattice contribution, $C_{lattice}$, and an electronic contribution, C_{el}, according to $C_p = C_{dil} + C_{lattice} + C_{el}$. Magnetic contributions are absent since ThOS is not magnetic. C_{dil}, according to the thermal lattice expansion, was assumed to be $C_{dil} = A \cdot C_p^2 \cdot T$ (with A = constant). $C_{lattice}$ was calculated from Debye functions, and the electronic contribution, $C_{el} = \gamma \cdot T$, was assumed to be zero. The constants $A(300\ K)$ and $\Theta_D(300\ K)$ were calculated to be $A(300) = 0.14 \times 10^{-5}$ and $\Theta_D(300) = 500\ K$ [9].

10.2.1.4 Chemical Reactions

With Elements

No reaction was observed with hydrogen up to red heat [1], see also [3]. ThOS is reported to react with carbon at about 1900°C forming ThS and CO [18], see also [2, 3]. But this reaction should lead to the formation of thorium carbide, too [2]. ThO_2 is formed if ThOS is heated up in an atmosphere of oxygen [1]. ThOS reacts with aluminium at about 1150°C to form ThS and Al_2O_3 [3], see also p. 3. The reaction of ThOS with thorium metal leads to the formation of ThS and ThO_2 [2].

With Compounds

ThOS dissolves very slowly in dilute acids and more rapidly in hot concentrated acids [2]. Slow dissolution is reported in HNO_3 and aqua regia [1]. ThOS reacts with H_2S (in the presence of carbon) to form ThS_2 (see p. 33) [4, 19], see also [2]. No formation of eutectic compositions was observed between ThOS and thorium sulfides [19]. Otherwise, polished sections of sintered ThS bodies with ThOS as a second phase showed that a ThOS-ThS eutectic liquid was formed in the temperature range 1900 to 2000°C [20].

References for 10.2:

[1] G. Krüss (Z. Anorg. Allgem. Chem. **6** [1894] 49/56). — [2] E. D. Eastman, L. Brewer, Le Roy A. Bromley, P. W. Gilles, L. Lofgren (J. Am. Chem. Soc. **73** [1951] 3896/8). — [3] J. Flahaut (Silicon Sulfur Phosphates Colloq., Münster, Ger., 1954 [1955], pp. 165/7; C.A. **1958** 3580). — [4] G. V. Samsonov, G. N. Dubrovskaya (Zh. Prikl. Khim. **36** [1963] 1615/8; J. Appl. Chem. [USSR] **36** [1963] 1555/7; C.A. **60** [1964] 8922). — [5] G. V. Samsonov, G. N. Dubrovskaya (At. Energ. [USSR] **15** [1963] 428/30; Soviet At. Energy **15** [1964] 1191/3; N.S.A. **18** [1964] No. 8342).

[6] H. Hartmann, D. Mootz, Th. Nentwich (Angew. Chem. **73** [1961] 172/3). — [7] L. Brewer, Le Roy A. Bromley, E. D. Eastman (U.S. Appl. 791466 [1950]; Offic. Gaz. U.S. Patent Office **641** [1950] 1346; C.A. **1952** 2258). — [8] R. Heindl, R. Loriers (Bull. Soc. Chim. France **1974** I 377/8). — [9] G. Amoretti, A. Blaise, J. M. Collard, R. O. A. Hall, M. J. Mortimer, R. Troc (J. Magn. Magn. Mater. **46** [1984] 57/67). — [10] G. Amoretti, D. C. Giori, V. Varacca, J. C. Spirlet, J. Rebizant (Phys. Rev. [3] B **20** [1979] 3573/7).

[11] H. U. Boelsterli, F. Hulliger (J. Mater. Sci. **3** [1968] 664/5). — [12] R. M. Dell, N. J. Bridger (MTP [Med. Tech. Publ. Co.] Intern. Rev. Sci. Inorg. Chem. Ser. One **7** [1972] 211/74). — [13] R. M. Dell, M. Allbutt (AERE-R-4253 [1963] 1/48; C.A. **59** [1963] 204). — [14] W. H. Zachariasen (Acta Cryst. **2** [1949] 291/6). — [15] L. I. Katzin (Natl. Nucl. Energy Ser. Div. IV B **14** [1954] 66/102).

[16] H. Matzke, R. Lindner (Atomkernenergie **9** [1964] 2/46). — [17] G. Amoretti, C. Fava, V. Varacca (Z. Naturforsch. **37a** [1982] 536/45). — [18] B. M. Abraham, N. R. Davidson (CN-

3001 [1945]) from E. D. Eastman, L. Brewer, Le Roy A. Bromley, P. W. Gilles, N. L. Lofgren (J. Am. Chem. Soc. **73** [1951] 3896/8, also mentioned in Natl. Nucl. Energy Ser. Div. IV B **14 I** [1949] 779/92). — [19] E. D. Eastman, L. Brewer, Le Roy A. Bromley, P. W. Gilles, N. L. Lofgren (J. Am. Chem. Soc. **72** [1950] 4019/23). — [20] P. D. Shalek (J. Am. Ceram. Soc. **46** [1963] 155/61).

10.3 Compounds of Thorium with Sulfur and Nitrogen

There is only one compound known within the system thorium-sulfur-nitrogen: Th_2N_2S.

10.3.1 Dithorium Dinitride Sulfide, Th_2N_2S

10.3.1.1 Preparation

Th_2N_2S was obtained by heating cold-pressed mixtures of ThS and ThN in proper amounts at 1500 to 1700 °C under 1 atm of N_2 for $1/2$ to 2 h in a tungsten crucible. The chemical analysis gave 87.7 $\pm$ 1 wt% Th, 4.7 $\pm$ 0.5 wt% N, 5.38 $\pm$ 0.1 wt% S, resulting in the formula $Th_2N_{1.8}S_{0.9}$ [1]. Th_2N_2S was also prepared by the reaction of ThN with pure sulfur in stoichiometric amounts sealed in an evacuated silica tube and heated to 1000 °C for 30 d [1], see also [2].

10.3.1.2 Crystallographic Properties

Th_2N_2S is hexagonal (Ce_2O_2S type) with one molecule per unit cell; the space group is $P\bar{3}m1$-D^3_{3d} (No. 164). The lattice parameters obtained from X-ray diffraction pattern are a = 4.008 $\pm$ 0.001 Å, c = 6.920 $\pm$ 0.002 Å [1], see also [2, 3]. The X-ray density was calculated from this data to be 9.04 g/cm^3. The measured pycnometric density is 8.1 g/cm^3 [1]. The atomic positions due to the space group $P\bar{3}m1$ are 2 Th in $\pm$ $(1/3, 2/3, u_1)$, 2 N in $\pm$ $(1/3, 2/3, u_2)$, 1 S in (0, 0, 0), with u_1 = 0.278 $\pm$ 0.005 (calculated from the observed intensities) and u_2 = 0.626 (if the nitrogen atoms are placed equidistantly from the four thorium atoms) [1], see also [3]. The interatomic distances are Th-4 N = 2.41, Th-3 S = 3.01 Å [1].

References for 10.3:

[1] R. Benz, W. H. Zachariasen (Acta Cryst. B **25** [1969] 294/6). — [2] R. M. Dell, N. J. Bridger (MTP [Med. Tech. Publ. Co.] Intern. Rev. Sci. Inorg. Chem. Ser. One **7** [1972] 211/74). — [3] D. J. Lam, J. B. Darby Jr., M. V. Nevitt (in: A. J. Freeman, J. B. Darby Jr., The Actinides: Electronic Structure and Related Properties, Vol. 2, Academic, New York 1974, pp. 119/84).

10.4. Compounds of Thorium with Rare Earth Elements and Sulfur

10.4.1 Phase Relationships

The existence of a range of ThS-CeS solid solutions with NaCl-type structure has been shown for compositions with 33 to 67% ThS [1], and also reported for $Sm_{1-x}Th_xS$ compounds with 5 to 10% Th [2]. A compound with composition $Th_{0.16}Y_{0.84}S_{1.3}$ has been reported to crystallize with NaCl-type structure [3].

Ln$_2^{III}$ThS$_5$ compounds have been prepared with Ln = La, Ce, Pr, Nd, and Sm and observed from X-ray diffraction pattern to crystallize in the orthorhombic system (U$_3$S$_5$ type) with the space group Pnma-D$_{2h}^{16}$ (No. 62) [4].

Ln$_4^{III}$Th$_5$S$_{16}$ compounds, prepared with Ln = Tb, Dy, Y, Ho, Er, Tm, Yb, and Lu, were shown to crystallize in the monoclinic system with the space group B2-C$_2^3$ (No. 5), Bm-C$_5^3$ (No. 8), or B2/m-C$_{2h}^3$ (No. 12) [5].

10.4.2 LnS-ThS Solid Solutions (Ln = Ce, Sm)

ThS and CeS form a range of solid solutions, Th$_x$Ce$_{1-x}$S, from which compositions with 33 to 67% ThS have been confirmed [1, 6]. Ce$_{1-x}$Th$_x$S compounds were prepared by heating either powdered mixtures of thorium hydride and Ce$_2$S$_3$ or ThS$_2$ and cerium hydride in proper amounts in vacuum, after pressing into pellets and placing in molybdenum crucibles [7].

The Ce$_{1-x}$Th$_x$S compounds crystallize with NaCl-type structure. Measured lattice parameters are listed in Table 32.

Table 32
Lattice Parameters of Ce$_{1-x}$Th$_x$S Compounds [6].

composition	a in kX
67% CeS, 33% ThS	5.725 ± 0.003
50% CeS, 50% ThS	5.709 ± 0.003
33% CeS, 67% ThS	5.701 ± 0.003

X-ray photoemission spectra of Sm$_{1-x}$Th$_x$S crystals were measured to provide information on the electronic structure and excitonic effects in these compounds (see **Fig. 23**) [2, 8, 9].

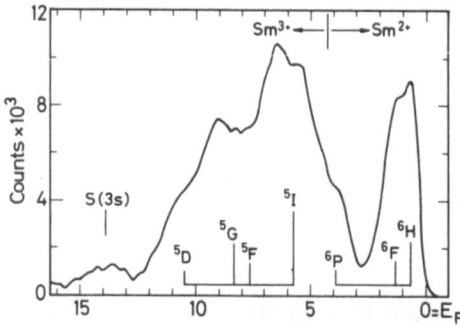

Fig. 23. XPS (X-ray photoemission spectrum) valence bands of Sm$_{0.85}$Th$_{0.15}$S [9].

Chemical Reactions

Sintered bodies of Ce$_{0.65}$Th$_{0.32}$S were observed to be excellently suitable as crucible material in the melting technology of plutonium [10].

10.4.3 $Y_{0.84}Th_{0.16}S_{1.3}$

$Th_{0.16}Y_{0.84}S_{1.3}$ is reported to have a NaCl-type structure with a lattice parameter of a = 5.529 ± 0.001 Å. At the very low temperature of T_C = 1.5 K this compound showed an inductive superconductivity transition. Superconducting transition temperatures were determined using low-frequency (22 Hz) induction techniques [3].

10.4.4 $Ln_2^{III}ThS_5$ Compounds (Ln = La, Ce, Pr, Nd, Sm)

Ln_2ThS_5 compounds were prepared by heating stoichiometric mixtures of ThS_2 and Ln_2S_3 at 1250°C in a stream of H_2S. They are orthorhombic (U_3S_5 type); the space group is Pnma-D_{2h}^{16} (No. 62) [4]. The lattice parameters, determined from X-ray powder diffraction patterns, are listed in Table 33.

Table 33
Lattice Parameters of Ln_2ThS_5 Compounds [4].

compound	a in Å	b in Å	c in Å	V in Å^3
La_2ThS_5	12.06	8.36	7.59	765
Ce_2ThS_5	12.01	8.29	7.56	753
Pr_2ThS_5	11.99	8.24	7.53	744
Nd_2ThS_5	11.96	8.21	7.50	736
Sm_2ThS_5	11.93	8.14	7.45	723

Magnetic susceptibility measurements were carried out at 4.2 to 300 K. At least in the high temperature region the Curie-Weiss law is obeyed (except for Sm_2ThS_3) [4]. The paramagnetic Curie temperatures, Θ, and the molar paramagnetic Curie constants, C_M, are listed in Table 34.

Table 34
Paramagnetic Properties of Ln_2ThS_5 Compounds [4].

	Ce_2ThS_5	Pr_2ThS_5	Nd_2ThS_5
Θ in K	−53	−22	−27
C_M	1.89	3.28	3.56
μ_{eff} of Ln^{3+} in μ_B	2.75	3.62	3.77

The reverse susceptibility of Sm_2ThS_5 does not obey the Curie-Weiss law, as previously observed for Sm_2ZrS_5. A maximum in the susceptibility curve was detected above 4.2 K [4].

References for 10.4 on p. 56

10.4.5   Ln$_4^{III}$ Th$_5$S$_{16}$ Compounds (Ln = Tb, Dy, Y, Ho, Er, Tm, Yb, Lu)

Ln$_4$Th$_5$S$_{16}$ compounds were prepared by two different methods. 1. Stoichiometric mixtures of powdered ThS$_2$ and Ln$_2$S$_3$, pressed into pellets (1 t/cm^2) and sealed in an evacuated silica tube, were heated at 1100°C for 24 h. 2. Stoichiometric mixtures of powdered ThO$_2$ and Ln$_2$O$_3$, placed in a carbon crucible, were heated in a stream of H$_2$S to 1350°C for 6 h. However, the first method was preferred [5].

The Ln$_4$Th$_5$S$_{16}$ compounds are monoclinic with four molecules per unit cell. The space group, determined by the systematic extinctions in Weissenberg pattern obtained from a single crystal of Tb$_4$Th$_5$S$_{16}$, is B2-C$_2^3$ (No. 5), Bm-C$_5^3$ (No. 8), or B2/m-C$_{2h}^3$ (No. 12) [5]. The lattice parameters are summarized in Table 35.

Table 35
Lattice Parameters of Ln$_4$Th$_5$S$_{16}$ Compounds [5].

compound	a in Å ±0.04	b in Å ±0.03	c in Å ±0.04	angle
Tb$_4$Th$_5$S$_{16}$	16.27	10.60	13.69	102°58′
Dy$_4$Th$_5$S$_{16}$	16.22	10.52	13.61	102°37′
Y$_4$Th$_5$S$_{16}$	16.16	10.46	13.51	102°12′
Ho$_4$Th$_5$S$_{16}$	16.10	10.39	13.43	101°90′
Er$_4$Th$_5$S$_{16}$	16.04	10.31	13.36	101°72′
Tm$_4$Th$_5$S$_{16}$	15.94	10.17	13.28	101°05′
Yb$_4$Th$_5$S$_{16}$	15.89	10.10	13.20	100°82′
Lu$_4$Th$_5$S$_{16}$	15.83	10.03	13.12	100°60′

10.4.6   CeThS$_2$

Sintered bodies of CeThS$_2$ were observed to be excellently suitable as crucible material in the melting technology of plutonium [10].

References for 10.4:

[1] E. D. Eastman, L. Brewer, Le Roy A. Bromley, P. W. Gilles, N. L. Lofgren (J. Am. Chem. Soc. **72** [1950] 4019/23). − [2] G. K. Wertheim, I. Nowik, M. Campagna (Z. Physik B **29** [1978] 193/7). − [3] A. R. Moodenbaugh, D. C. Johnston, R. Viswanathan, R. N. Shelton, L. E. de Long, W. A. Fertig (J. Low Temp. Phys. **33** [1978] 175/203). − [4] H. Noël, J. Prigent (Physica B + C **102** [1980] 372/9). − [5] Vo-Van Tein, M. Guittard (Compt. Rend. C **279** [1974] 849/50).
[6] W. H. Zachariasen (Acta Cryst. **2** [1949] 291/6). − [7] L. Brewer, Le Roy A. Bromley (AECD-2242 [1948] 1/47; N.S.A. **1** [1948] No. 1098). − [8] M. Campagna, E. Bucher, G. K. Wertheim, L. D. Longinotti (Phys. Rev. Letters **33** [1974] 165/8). − [9] M. Campagna, E. Bucher, G. K. Wertheim, L. D. Longinotti (Proc. 11th Rare Earth Res. Conf., Traverse City, Mich., 1974, Vol. 1, pp. 53/62). − [10] S. Fried, E. F. Westrum Jr., H. L. Baumbach, P. L. Kirk (J. Inorg. Nucl. Chem. **5** [1958] 182/9).

10.5 Thorium Sulfite Compounds

David Brown
Chemistry Division, A.E.R.E.
Harwell, Oxon, England

10.5.1 Introduction

Thorium sulfites have not been extensively studied and few compounds have been characterised. The pre-1950 literature on phases of composition $Th(SO_3)_2 \cdot n H_2O$ (n = 1, 4) and $Th_2(OH)_2(SO_3)_3 \cdot 37 H_2O$ is covered in "Thorium" 1955, p. 283, and that on complexes such as $(NH_4)_4Th(SO_3)_4 \cdot 6 H_2O$, $Na_4Th_2(OH)_2(SO_3)_5 \cdot n H_2O$ (n = 12 to 22) and $K_4Th(OH)_2(SO_3)_3 \cdot 10 H_2O$ on pp. 325, 333, and 338. The sulfites, sulfito complexes, and basic sulfites reported to-date are listed in Table 36. It should be noted that additional studies are required to confirm the existence of many of the basic sulfites since it has not always been established that the products were single phase. In addition to the compounds listed a few mixed acid sulfito complexes (Chapter 10.5.7, p. 62) and basic sulfito complexes (Chapter 10.5.6, p. 62) have been described.

Table 36
Thorium Sulfites, Basic Sulfites, and Sulfito Complexes.

sulfites	basic sulfites [1]	sulfito complexes
$Th(SO_3)_2 \cdot 4 H_2O$	$Th_2O(SO_3)_3 \cdot n H_2O$ (n = 2 and 0)	$M_2^ITh(SO_3)_3 \cdot n H_2O$ (M^I = Na, n = 5 and 2;
$Th(SO_3)_2 \cdot 2 H_2O$		K, n = 7.5 and 2; NH_4,
	$Th_2(OH)_2(SO_3)_3 \cdot n H_2O$	n = 4 and 2; CN_3H_6,
$Th(SO_3)_2$?	(n = 11, 7, 3, and 0)	n = 12, 10, 2, 0)
	$ThO(SO_3)$	$Na_6Th_2(SO_3)_7 \cdot 18 H_2O$
	$Th_2O(OH)_2(SO_3)_2 \cdot n H_2O$ (n = 8 to 10)	$M_4^ITh(SO_3)_4 \cdot n H_2O$ (M^I = Na, x = 6, 3 and 1; NH_4, n = 5 and 6)
	$Th_2(OH)_4(SO_3)_2 \cdot 3 H_2O$	
	$Th_2O_2(OH)_2(SO_3)$	
	$Th_2(OH)_6(SO_3) \cdot 13 H_2O$	

[1] Additional studies are required to confirm the existence of many of the basic sulfites.

10.5.2 Thorium(IV) Sulfites, $Th(SO_3)_2 \cdot 4 H_2O$, $Th(SO_3)_2 \cdot 2 H_2O$, $(Th(SO_3)_2?)$

Preparation

The addition of Na_2SO_3 or SO_2 to a thorium nitrate solution at room temperature gives $Th(SO_3)_2 \cdot 4 H_2O$ [1]. The precipitation with SO_2 is done at thorium nitrate concentrations below ca. 0.3 M. With higher concentrations it is necessary to dilute the yellow-green solution to

References for 10.5.2 on p. 59

induce precipitation. Partially hydrolysed products are formed when the reaction is performed at 60 to 85 °C [1]. Alternatively the reaction of $ThOCl_2 \cdot n\, H_2O$ with SO_2 has been employed for the preparation of $Th(SO_3)_2 \cdot 4\, H_2O$ [5]. The tetrahydrate loses two molecules of water in the temperature range 75 to 130 °C to yield the dihydrate, $Th(SO_3)_2 \cdot 2\, H_2O$, although this appears to have been identified only by the weight loss [1]. Further loss of water between 135 and 150 °C was originally reported to result in the formation of the anhydrous bis(sulfite) [1]. However, later IR studies [2, 3] on the product obtained at ca. 200 °C indicated that partial oxidation to sulfate had occurred together with loss of water and that the pure bis(sulfite) had not been obtained. Despite this the same authors later reported that $Th(SO_3)_2$ is obtained by heating $Th(SO_3)_2 \cdot 1.5\,(CH_3)_2CO \cdot 1.5\, H_2O$ at 140 to 215 °C and that in this case oxidation to $Th(SO_4)_2$ was observed at 280 to 300 °C [4].

Properties

X-ray structural data are not available for either of the bis(sulfite) hydrates. Infrared results for $Th(SO_3)_2 \cdot 4\, H_2O$ (in cm^{-1}: ν_{OH}, 2800 to 3670; ν_{H_2O}, 1642; H_2O twist, 545 cm^{-1}; $\nu_{SO_3^{2-}}$ at 470, 633, 850, 922, and 1000) have been interpreted on the basis of bidentate or bridging sulfito groups with bonding intermediate between ionic and that typical of coordinated sulfite, but closer to ionic. The water molecules are involved in hydrogen bonding [2].

The refractive index of $Th(SO_3)_2 \cdot 4\, H_2O$, determined by immersion, is 1.607; the pycnometric density is 3.10 g/cm^3 [1].

The estimated enthalpy of formation for $Th(SO_3)_2$ is -490.2 kcal/mol [6] and it is calculated [7] that the dissociation pressure is given by the equation, $\log p_{Torr}\,(SO_2) = -6140/T + 11.154$.

The conversion of the tetrahydrate to the dihydrate and impure $Th(SO_3)_2$ on heating in air is mentioned above. According to an IR study of products formed at temperatures between 200 and 650 °C oxidation to $Th(SO_4)_2$ occurs in this temperature range, with an exothermic effect at 400 to 450 °C on the thermogram being associated with oxidation of the greater part of the sulfite and one at 650 °C corresponding to conversion of amorphous $Th(SO_4)_2$ into one of its crystal modifications. The thermal decomposition of a sample of $Th(SO_3)_2 \cdot 4\, H_2O$ prepared at 80 °C (analyses corresponding to $ThO_{0.25}(SO_3)_{1.75} \cdot 4\, H_2O$; $D_{pycn} = 3.85$ g/cm^3) has been reported [1] to give the dihydrate (75 to 130 °C), $Th(SO_3)_2$ (135 to 150 °C), and $ThO(SO_3)$ (700 to 715 °C) but, in view of the above IR results indicating sulfate formation [2, 3] on decomposition of pure $Th(SO_3)_2 \cdot 4\, H_2O$, the formation of $Th(SO_3)_2$ and $ThO(SO_3)$, based only on weight changes, cannot be accepted as established. The ultimate decomposition product, at 720 to 800 °C, is ThO_2.

$Th(SO_3)_4 \cdot 4\, H_2O$ is readily decomposed by mineral acids with vigorous evolution of SO_2 [1]. It undergoes partial hydrolysis in air to give, after one month at 100% relative humidity, a colourless glassy solid of composition $Th_2(OH)_2(SO_3)_3 \cdot 11\, H_2O$ [1]. This conversion also occurs to a significant extent over H_2SO_4 [3].

The hydrated acetone adduct $Th(SO_3)_2 \cdot 1.5\,(CH_3)_2CO \cdot 1.5\, H_2O$ is formed when thorium(IV) hydroxide suspended in acetone is treated with SO_2. However, when freshly prepared $Th(SO_3)_2 \cdot 4\, H_2O$ suspended in a water/acetone mixture is treated in a similar way the product is reported to be $Th_2O(SO_3)_3 \cdot (CH_3)_2CO \cdot 7\, H_2O$ [4]. The reaction between $Th(SO_3)_2 \cdot 4\, H_2O$ and molten acetamide gives, depending on the reaction period and temperature, either $Th_2(OH)_2(SO_3)_3 \cdot 2\, CH_3CONH_2 \cdot 4\, H_2O$ or $Th_2(OH)_4(SO_3)_2 \cdot CH_3CONH_2 \cdot 5\, H_2O$ [4].

References for 10.5.2:

[1] V. A. Golovnya, A. D. Molodkin, V. N. Tverdokhlebov (Zh. Neorgan. Khim. **9** [1964] 2032/4; Russ. J. Inorg. Chem. **9** [1964] 1097/8). — [2] Yu. Ya. Kharitonov, V. N. Tverdokhlebov, A. K. Molodkin (Zh. Neorgan. Khim. **11** [1966] 2608/13; Russ. J. Inorg. Chem. **11** [1966] 1401/4). — [3] V. A. Golovnya, A. K. Molodkin, V. N. Tverdokhlebov (Zh. Neorgan. Khim. **12** [1967] 2377/87; Russ. J. Inorg. Chem. **12** [1967] 1254/9). — [4] V. A. Golovnya, A. K. Molodkin, V. N. Tverdokhlebov (Zh. Neorgan. Khim. **12** [1967] 2729/39; Russ. J. Inorg. Chem. **12** [1967] 1439/44). — [5] V. A. Golovnya, A. K. Molodkin, V. N. Tverdokhlebov (Zh. Neorgan. Khim. **12** [1967] 2075/85; Russ. J. Inorg. Chem. **12** [1967] 1092/8).

[6] E. Erdos (Collection Czech. Chem. Commun. **27** [1962] 1428/37). — [7] E. Erdos (Collection Czech. Chem. Commun. **27** [1962] 2273/83).

10.5.3 Thorium(IV) Basic Sulfites

Preparation

Hydrolysis of $Th(SO_3)_2 \cdot 4\,H_2O$ (p. 57) in moist air [1, 2] and the action of SO_2 on aqueous solutions of $Th(OH)_3Cl$ [2] both give $Th_2(OH)_2(SO_3)_3 \cdot 11\,H_2O$. Two further basic sulfites have been reported to precipitate from ethereal solutions. Thus, products with the compositions $Th_2O(OH)_2(SO_3)_2 \cdot n\,H_2O$ and $Th_2(OH)_6SO_3 \cdot 13\,H_2O$ have been obtained [2] by bubbling SO_2 through suspensions of thorium hydroxide and hydrated thorium dihydroxide carbonate, respectively. The water content of the former ($= n$) apparently varies between 8 and 10 depending on the extent of hydration of the hydroxide starting material.

Additional phases have been reported on the basis of DTA and TGA studies. For example, thermal dehydration of $Th_2(OH)_2(SO_3)_3 \cdot 11\,H_2O$ is stated to proceed via the heptahydrate (60 to 90 °C) and the trihydrate (90 to 135 °C) to the anhydrous sulfite, $Th_2(OH)_2(SO_3)_3$, at 340 to 370 °C; $Th_2O(SO_3)_3$ is reported to form at 700 °C and this is converted to ThO_2 at 750 to 800 °C [1]. Decomposition of a partially hydrolysed sample of $Th(SO_3)_2 \cdot 4\,H_2O$ (possible composition given as $ThO_{0.25}(SO_3)_{1.75} \cdot 4\,H_2O$) obtained by the action of SO_2 on an aqueous solution of $Th(NO_3)_4$ at 80 °C is claimed to give $ThO(SO_3)$ at 700 to 715 °C [1]. However, these interpretations of the thermograms and mass-loss curves are probably not entirely satisfactory in view of the reported simultaneous partial oxidation of sulfite to sulfate on dehydration of $Th(SO_3)_2 \cdot 4\,H_2O$ below 200 °C [3, 4]. Thus the existence of phases such as $Th_2(OH)_2(SO_3)_3$, $Th_2O(SO_3)_3$, and $ThO(SO_3)_3$ requires confirmation.

Similarly, the reported formation of $Th_2O(SO_3)_3 \cdot 2\,H_2O$, $Th_2(OH)_4(SO_3)_2 \cdot 3\,H_2O$, and possibly $Th_2O_2(OH)_2(SO_3)$ as intermediates in the thermal decomposition of the oxygen donor adducts $Th_2O(SO_3)_3 \cdot (CH_3)_2CO \cdot 7\,H_2O$ (at 100 to 130 °C), $Th_2(OH)_4(SO_3)_2 \cdot CH_3CONH_2 \cdot 5\,H_2O$ (at 240 to 330 °C), and $Th_2(OH)_2(SO_3)_3 \cdot 2\,CH_3CONH_2 \cdot 5\,H_2O$ (at 315 to 550 °C), respectively, is based on DTA and TGA studies [5], and additional work is needed to establish that these are genuine phases and not mixtures.

Properties

The densities reported for $Th_2(OH)_2(SO_3)_3 \cdot 11\,H_2O$, $Th_2O(OH)_2(SO_3)_2 \cdot 8\,H_2O$, and $Th_2(OH)_6(SO_3) \cdot 13\,H_2O$ are 3.15, 4.0 and 4.34 g/cm³, respectively. The refractive indices of the first two of these compounds are 1.550 and 1.651, respectively.

The preparation of oxide- and hydroxide sulfites containing donor ligands (viz. $Th_2O(SO_3)_3 \cdot (CH_3)_2CO \cdot 7\,H_2O$, $Th_2(OH)_2(SO_3)_3 \cdot 2\,CH_3CONH_2 \cdot 4\,H_2O$, and $Th_2(OH)_4(SO_3)_2 \cdot CH_3CONH_2 \cdot 5\,H_2O$) is dealt with on p. 58.

References for 10.5.3 on p. 60

References for 10.5.3:

[1] V. A. Golovnya, A. K. Molodkin, V. N. Tverdokhlebov (Zh. Neorgan. Khim. **9** [1964] 2032/4; Russ. J. Inorg. Chem. **9** [1964] 1097/8). — [2] V. A. Golovnya, A. K. Molodkin, V. N. Tverdokhlebov (Zh. Neorgan. Khim. **12** [1967] 2075/85; Russ. J. Inorg. Chem. **12** [1967] 1092/8). — [3] Yu. Ya. Kharitonov, V. N. Tverdokhlebov, A. K. Molodkin (Zh. Neorgan. Khim. **11** [1966] 2608/13; Russ. J. Inorg. Chem. **11** [1966] 1401/4). — [4] V. A. Golovnya, A. K. Molodkin, V. N. Tverdokhlebov (Zh. Neorgan. Khim. **12** [1967] 2377/87; Russ. J. Inorg. Chem. **12** [1967] 1254/9). — [5] V. A. Golovnya, A. K. Molodkin, V. N. Tverdokhlebov (Zh. Neorgan. Khim. **12** [1967] 2729/39; Russ. J. Inorg. Chem. **12** [1967] 1439/44).

10.5.4   Thorium(IV) Mixed Acid Sulfites

According to Golovnya et al. [1] $Th(OH)(SO_3)Cl$ and $Th_2O(SO_3)_2Cl_2$ are formed when the acetone adduct $Th(OH)(SO_3)Cl \cdot 0.5 (CH_3)_2CO \cdot 3 H_2O$ is heated in the temperature ranges 140 to 150°C and 150 to 500°C, respectively. It is surprising, in view of other data on the decomposition of thorium sulfites, that the sulfite ion should be present at temperatures as high as 500°C; oxidation to sulfate would be more probable.

Reference for 10.5.4:

[1] V. A. Golovnya, A. K. Molodkin, V. N. Tverdokhlebov (Zh. Neorgan. Khim. **12** [1967] 2729/39; Russ. J. Inorg. Chem. **12** [1967] 1439/44).

10.5.5   Thorium(IV) Sulfito Complexes

Preparation

Tris(sulfito) complexes of the type $M_2Th(SO_3)_3 \cdot x H_2O$ (M = Na, K, NH_4, and CN_3H_6 with x = 5, 7.5, 4 and 12, respectively) have been prepared by passing SO_2 through aqueous solutions containing $Th(SO_3)_2$ hydrate and the appropriate alkali, etc. sulfite. The products crystallised on cooling a previously heated mixture (60°C) to 16 to 18°C; the pH ranged from 4.6 to 6.5 [1]. The ammonium salt may also be prepared by passing SO_2 through an $(NH_4)_2CO_3$ solution in which thorium hydroxide has been dissolved [2]. Either ethanol or p-phenylenediamine was employed as anti-oxidant.

Dihydrates of the tris(sulfito) complexes are formed when the higher hydrates are heated at moderate temperatures (viz. Na and K, 50 to 180°C; NH_4, 30 to 135°C; CN_3H_6, 180 to 200°C). Decomposition of $(CN_3H_6)_2Th(SO_3)_3 \cdot 12 H_2O$ appears to proceed via the intermediate formation of a decahydrate at 50 to 120°C. Further water is lost from $(CN_3H_6)_2Th(SO_3)_3 \cdot 2 H_2O$ with the formation of the anhydrous salt at 200 to 240°C; this is apparently converted to $(CN_3H_6)_2Th_2(SO_3)_5$ at 240 to 400°C [3].

The tetrakis(sulfito) complexes $Na_4Th(SO_3)_4 \cdot 6 H_2O$ and $(NH_4)_4Th(SO_3)_4 \cdot 5 H_2O$ may also be isolated from solutions containing a larger excess of the sodium or ammonium sulfite [1]. A different hydrate with sodium as the cation, $Na_4Th(SO_3)_4 \cdot 3 H_2O$, is reported to form when the initial precipitate from an aqueous ethanol solution containing $Th(NO_3)_4$ and $Na_2(SO_3)$ (1:4 mole ratio) is washed extensively with ethanol [4]. Attempts to prepare higher sulfito complexes such as $Na_6Th(SO_3)_5 \cdot x H_2O$, $Na_8Th(SO_3)_6 \cdot x H_2O$, and $Na_{12}Th(SO_3)_8 \cdot x H_2O$ by increasing the Na_2SO_3 to $Th(NO_3)_4$ ratio gave irreproducible results [4].

The tetrakis complex, $Na_4Th(SO_3)_4 \cdot 6\,H_2O$, is converted into a trihydrate in the temperature range 100 to 160°C and to a monohydrate between 160 and 200°C [3]. The latter apparently also forms when $Na_4Th(SO_3)_4 \cdot 0.5\,CO(NH_2)_2 \cdot 6\,H_2O$ is heated in the temperature range 225 to 285°C [5].

According to [1], by increasing the pH from the range 5.8 to 6.0 to between 6.1 and 6.5 the salt $Na_6Th_2(SO_3)_7 \cdot 18\,H_2O$ is formed instead of $Na_2Th(SO_3)_3 \cdot 5\,H_2O$.

Properties

Infrared spectra are illustrated and band positions listed, without assignments, in reference [3] for the white complexes $M_2Th(SO_3)_3 \cdot x\,H_2O$ (M = Na, K, NH_4, and CN_3H_6 with x = 5, 7.5, 4, and 12, respectively), $Na_6Th_2(SO_3)_7 \cdot 18\,H_2O$, and $M_4Th(SO_3)_4 \cdot x\,H_2O$ (M = Na and NH_4 with x = 6 and 5, respectively). Infrared data are also provided for products formed when these complexes are heated at various temperatures up to 450°C. The results combined with NMR data (two signals for $K_2Th(SO_3)_3 \cdot 7.5\,H_2O$ at 2.17 and 0.51 Gauss, respectively) and thermograms are claimed to indicate that the tris- and tetrakis(sulfito) complexes have, respectively, two water molecules and one water molecule in the inner co-ordination sphere. There are no X-ray structural data available to substantiate this.

Refractive indices and pycnometrically determined densities are listed in Table 37 [1].

The conversion of the higher to lower hydrates is discussed above (p. 60/1). The ultimate decomposition products of the sodium salts and the potassium salt, at ca. 600 to 800°C, are mixtures of the alkali metal sulfate and ThO_2; the ammonium and guanidinium complexes are converted to ThO_2 via the bis(sulfate) in the temperature range 460 to 830°C [3].

The following complexes with urea have been reported: $Na_4Th(SO_3)_4 \cdot 0.5\,CO(NH_2)_2 \cdot 6\,H_2O$, $Na_6Th_2(SO_3)_7 \cdot 2\,CO(NH_2)_2 \cdot 15\,H_2O$, and $Na_{18}Th_4(SO_3)_{17} \cdot CO(NH_2)_2 \cdot 18\,H_2O$ [2, 5].

Table 37
Densities and Refractive Indices for Thorium(IV) Sulfito Complexes [1].

compound	density in g/cm³	refractive index
$Na_2Th(SO_3)_3 \cdot 5\,H_2O$	2.962	—
$K_2Th(SO_3)_3 \cdot 7.5\,H_2O$	2.976	—
$(NH_4)_2Th(SO_3)_3 \cdot 4\,H_2O$	2.996	1.591
$(CN_3H_6)_2Th(SO_3)_3 \cdot 12\,H_2O$	2.482	1.594
$Na_6Th_2(SO_3)_7 \cdot 18\,H_2O$	3.369	—
$Na_4Th(SO_3)_4 \cdot 6\,H_2O$	2.889 [1)]	1.552
$(NH_4)_4Th(SO_3)_4 \cdot 5\,H_2O$	2.595	1.558

[1)] Given in [3] as 2.735 g/cm³.

References for 10.5.5:

[1] V. A. Golovnya, A. K. Molodkin, V. N. Tverdokhlebov (Zh. Neorgan. Khim. **12** [1967] 2075/85; Russ. J. Inorg. Chem. **12** [1967] 1092/8). — [2] V. A. Golovnya, A. K. Molodkin, V. N.

Tverdokhlebov (Zh. Neorgan. Khim. **10** [1965] 2196/8; Russ. J. Inorg. Chem. **10** [1965] 1195/6).
– [3] V. A. Golovnya, A. K. Molodkin, V. N. Tverdokhlebov (Zh. Neorgan. Khim. **12** [1967] 2377/
87; Russ. J. Inorg. Chem. **12** [1967] 1254/9). – [4] L. N. Essen, D. P. Alekseeva, A. D. Gel'man
(Zh. Neorgan. Khim. **11** [1966] 1596/604; Russ. J. Inorg. Chem. **11** [1966] 853/7). – [5] V. A.
Golovnya, A. K. Molodkin, V. N. Tverdokhlebov'(Zh. Neorgan. Khim. **12** [1967] 2729/39; Russ.
J. Inorg. Chem. **12** [1967] 1439/44).

10.5.6   Thorium(IV) Hydroxo Sulfito Complexes

Attempts to prepare a sulfitothorate(IV) containing the $[Co(NH_3)_6]^{3+}$ cation have given only
a range of impure products, believed to be partially hydrolysed tetrakis(sulfito) species [1].

The formation of $(NH_4)_3Th(OH)(SO_3)(SO_4)_2 \cdot 3 H_2O$ as a consequence of oxidation of sulfite
to sulfate during the preparation of sulfito complexes is mentioned in [2] but no preparative
details or properties are given.

$Na_6Th_2(OH)_8(SO_3)_3$ is reported to form when the urea adduct $Na_6Th_2(OH)_8(SO_3)_3 \cdot 2 CO(NH_2)_2$
$\cdot 20 H_2O$ is heated at 120 to 150°C. It decomposes to a mixture of Na_2SO_4 and ThO_2 between
470 and 550°C [3].

References for 10.5.6:

[1] M. Hoshi, K. Ueno (J. Nucl. Sci. Technol [Tokyo] **15** [1978] 141/4). – [2] V. A. Golovnya,
A. K. Molodkin, V. N. Tverdokhlebov (Zh. Neorgan. Khim. **10** [1965] 2196/8; Russ. J. Inorg.
Chem. **10** [1965] 1195/6). – [3] V. A. Golovnya, A. K. Molodkin, V. N. Tverdokhlebov (Zh.
Neorgan. Khim. **12** [1967] 2729/39; Russ. J. Inorg. Chem. **12** [1967] 1439/44).

10.5.7   Thorium(IV) Mixed Acid Sulfito Complexes

Reactions in aqueous solution involving varying ratios of $Th(C_2O_4)_2 \cdot 6 H_2O$ and Na_2SO_3
are reported to give a series of mixed oxalato sulfito complexes of general formula
$Na_{2n}Th(SO_3)_n(C_2O_4)_2 \cdot x H_2O$ (n = 3, 4, 5, 6, 7, and 9 with x = 6, 6, 6, 8, 5, and 6, respectively)
[1, 2]. Treatment of $Na_{12}Th(SO_3)_6(C_2O_4)_2 \cdot 8 H_2O$ in water with $Ba(NO_3)_2$ results in the immediate
precipitation of $BaTh(SO_3)_6(C_2O_4)_2 \cdot 7 H_2O$ [2]. These unusual complexes require further
investigation.

According to Golovnya et al. [3] oxidation of sulfite during the preparation of sulfito
complexes can lead to the formation of sulfato sulfito complexes. The following phases are
reported but no preparative details or properties are given: $Na_4Th(SO_3)_{3.5}(SO_4)_{0.5} \cdot 7.5 H_2O$,
$Na_6Th(SO_3)_{3.5}(SO_4)_{1.5} \cdot 9 H_2O$, and $(NH_4)_4H_4Th(SO_3)_5(SO_4) \cdot 12 H_2O$.

References for 10.5.7:

[1] A. D. Gel'man, L. N. Essen, F. A. Zakharova, D. P. Alekseeva, M. M. Orlova (Dokl. Akad.
Nauk SSSR **149** [1963] 1071/3; Dokl. Chem. Proc. Acad. Sci. USSR **148/153** [1963] 317/9). –
[2] L. N. Essen, D. P. Alekseeva, A. D. Gel'man (Zh. Neorgan. Khim. **11** [1966] 1596/604; Russ.
J. Inorg. Chem. **11** [1966] 853/7). – [3] V. A. Golovnya, A. K. Molodkin, V. N. Tverdokhlebov
(Zh. Neorgan. Khim. **10** [1965] 2196/8; Russ. J. Inorg. Chem. **10** [1965] 1195/6).

10.6 Thorium Sulfate Compounds

David Brown
Chemistry Division, A.E.R.E.
Harwell, Oxon, England

10.6.1 Introduction

Thorium bis(sulfate) hydrates and oxide sulfate hydrates reported to-date are listed in Table 38 (p. 64) with the anhydrous compounds and the presently known peroxide sulfate hydrate and dihydroxide sulfate. The pre-1950 literature is discussed in "Thorium" 1955, pp. 283/93. There have been no recent reports concerning the existence of the bis(sulfate) hexahydrate, $Th(SO_4)_2 \cdot 6 H_2O$ ("Thorium" 1955, p. 288) and although the dihydrate ("Thorium" 1955, p. 289) and monohydrate have been reported as intermediates in the thermal decomposition of higher hydrates they were not isolated and studied. Thus, the only well-characterised bis(sulfate) hydrates are $Th(SO_4)_2 \cdot 9 H_2O$, $Th(SO_4)_2 \cdot 8 H_2O$, and $Th(SO_4)_2 \cdot 4 H_2O$. Similarly, there have been no further publications dealing with the oxide sulfate hydrates $ThOSO_4 \cdot n H_2O$ (n = 5, 3, 2 and 1; "Thorium" 1955, pp. 292/3) and it is not yet established whether these were genuine phases or hydrates of, or in the case of n = 1 the same phase as the well-characterised dihydroxide sulfate, $Th(OH)_2SO_4$.

The reported sulfato complexes are listed in Table 39 (p. 64). The pre-1950 publications are covered in "Thorium" 1955, pp. 325, 333/4, 338/9, 343, and 344/6. Further work is required to confirm the existence of the more unusual phases such as, for example, $Na_2Th(SO_4)_3 \cdot 12 H_2O$, $SnTh(SO_4)_4 \cdot 4 H_2O$, $K_{14}Th_2(SO_4)_{11}$, and $Tl_{14}Th_2(SO_4)_{11}$.

10.6.2 Thorium Bis(sulfate), $Th(SO_4)_2$

Preparation

Anhydrous thorium bis(sulfate), $Th(SO_4)_2$, is conveniently prepared by the thermal decomposition of hydrated bis(sulfates) at 300 to 450°C [1 to 3] or of ammonium sulfatothorate(IV) hydrates of the types $(NH_4)_{2m-4}[Th(SO_4)_m] \cdot n H_2O$ (m = 3, 4, 5, and 6 with n = 5, 2, 3, and 2, respectively) at ca. 550°C [2, 3]. Removal of hydrazine from the complex $Th(SO_4)_2 \cdot 2 N_2H_4$ (see "Thorium" E, 1985, p. 9) at 265°C also gives $Th(SO_4)_2$ [4]. Other less direct preparations involve thermal decomposition of the bis(sulfite) hydrates [5, 6], $(NH_4)_2Th(SO_3)_3 \cdot 4 H_2O$ [6], $(CN_3H_6)_2Th(SO_3)_3 \cdot 12 H_2O$ [6], and $Th_2(SO_3)_4 \cdot (CH_3)_2CO \cdot 3 H_2O$ [7], and also of $Th(SO_3F)_4$ [8].

Physical Properties

The structure of $Th(SO_4)_2$, which is a white solid, is unknown. According to thermal analysis data there is no phase change in the temperature range 298 to 900 K [9].

The early measurements on the enthalpy of solution of $Th(SO_4)_2$ in NaOH of undefined concentration [10] and on the decomposition pressures of $Th(SO_4)_2$ [11] are now considered to be unreliable. The results reported by Mayer et al. [9] for manometrically determined equilibrium decomposition pressures of the reaction $Th(SO_4)_2(c) \rightarrow ThO_2(c) + 2 SO_2(g) + O_2$ (908 to 1057 K) and their enthalpy increments for anhydrous $Th(SO_4)_2$ measured by drop calorimetry in the temperature range 623 to 897 K have been evaluated by Rand [12] and

References for 10.6.2 on pp. 66/7

Table 38
Thorium(IV) Sulfates.

$Th(SO_4)_2$	$ThOSO_4$
$Th(SO_4)_2 \cdot n\,H_2O$ (n = 9, 8, 6, 4, 2, and 1)[1]	$ThOSO_4 \cdot n\,H_2O$ (n = 5, 3, 2, and 1) $Th(O_2)SO_4 \cdot 3\,H_2O$ $Th(OH)_2SO_4$

[1] There have been no recent publications concerning $Th(SO_4)_2 \cdot 6\,H_2O$ and although $Th(SO_4)_2 \cdot 2\,H_2O$ and $Th(SO_4)_2 \cdot H_2O$ have been reported as intermediates in the thermal decomposition of higher hydrates (p. 69) they were not isolated and characterised.

Table 39
Thorium(IV) Sulfato Complexes. Complexes marked with an asterisk have not been studied since publication of Main Volume "Thorium", 1955. The few reported mixed acid sulfato complexes are discussed in Chapter 10.6.9 (p. 87).

tris(sulfato) complexes	tetrakis(sulfato) complexes	pentakis(sulfato) complexes	hexakis(sulfato) complexes	miscellaneous sulfato complexes
$Na_2Th(SO_4)_3 \cdot n\,H_2O$ (n = 12*, 6, 4, and 0)	$Na_4Th(SO_4)_4 \cdot 4\,H_2O$*	$K_6Th(SO_4)_5 \cdot 3\,H_2O$	$K_8Th(SO_4)_6 \cdot 2\,H_2O$	$Na_{10}Th_2(SO_4)_9 \cdot n\,H_2O$ (n = 5 and 0)
$K_2Th(SO_4)_3 \cdot n\,H_2O$ (n = 4 and 0)	$K_4Th(SO_4)_4 \cdot n\,H_2O$ (n = 2 and 0)	$Cs_6Th(SO_4)_5 \cdot n\,H_2O$ (n = 3 and 0)	$(NH_4)_8Th(SO_4)_6 \cdot n\,H_2O$ (n = 2 and 0)	$K_{14}Th_2(SO_4)_{11}$*
$Rb_2Th(SO_4)_3 \cdot n\,H_2O$ (n = 2 and 0)	$Rb_4Th(SO_4)_4$	$(NH_4)_6Th(SO_4)_5 \cdot n\,H_2O$ (n = 3 and 0)		$Tl_{14}Th_2(SO_4)_{11}$*
$Cs_2Th(SO_4)_3 \cdot n\,H_2O$ (n = 2 and 0)	$Cs_4Th(SO_4)_4 \cdot n\,H_2O$ (n = 1 and 0)	$[Co(NH_3)_6]_2Th(SO_4)_5 \cdot 2\,H_2O$		
$Tl_2Th(SO_4)_3 \cdot 4\,H_2O$*	$(NH_4)_4Th(SO_4)_4 \cdot n\,H_2O$ (n = 2 and 0)			
$(NH_4)_2Th(SO_4)_3 \cdot n\,H_2O$ (n = 5, 4*, and 0)	$(C_6H_5N_2H_4)_4Th(SO_4)_4$*			
$(C_2H_5NH_3)_2Th(SO_4)_3 \cdot 4\,H_2O$*	$(C_8H_{17}NH_3)_4Th(SO_4)_4$*			
$[(C_2H_5)_2NH_2]_2Th(SO_4)_3 \cdot 4\,H_2O$*	$SnTh(SO_4)_4 \cdot 2\,H_2O$*			
$(C_5H_6N)_2Th(SO_4)_3 \cdot 4\,H_2O$*				
$(C_9H_8N)_2Th(SO_4)_3 \cdot 4\,H_2O$*				
$(CN_3H_6)NaTh(SO_4)_3 \cdot n\,H_2O$ (n = 6 and 0)				

Table 40
Thermodynamic Quantities for Thorium Sulfate, $Th(SO_4)_2$ [13].
Thermodynamic quantities for thorium sulfate in SI units are also given in [13].

T in K	C_p° in cal·mol⁻¹·K⁻¹	S° in cal·mol⁻¹·K⁻¹	$-(G_T^\circ - H_{298}^\circ)/T$ in cal·mol⁻¹·K⁻¹	$H_T^\circ - H_{298}^\circ$ in cal/mol	ΔH_f° in cal/mol	ΔG_f° in cal/mol	log K_p
0					−594 138		
298	41.458	40.000	40.000	0	−606 100	−551 229	404.056
300	41.560	40.257	40.001	77	−606 107	−550 889	401.317
400	47.080	52.969	41.697	4 509	−607 382	−532 400	290.886
500	52.600	64.067	45.082	9 493	−607 885	−513 585	224.485
600	58.120	74.145	49.097	15 029	−607 790	−494 724	180.201
700	63.640	83.519	53.352	21 117	−607 141	−475 875	148.573
800	69.160	92.377	57.682	27 757	−632 089	−459 736	125.592
900	74.680	100.842	62.010	34 949	−629 795	−438 323	106.438
1000	80.200	108.996	66.303	42 693	−627 027	−417 189	91.175

References for 10.6.2 on pp. 66/7

5

subsequently by Cordfunke and O'Hare [13]. The following values, of which the standard entropy is estimated, are recommended [13]: $\Delta H_f^\circ[Th(SO_4)_2, c, 298\ K] = -606.1 \pm 5.0$ kcal/mol; C_p (in cal $\cdot$ mol^{-1} $\cdot$ K^{-1}) $= 25.0 + (55.2 \times 10^{-3})T$ (600 to 900 K); $\Delta G_f^\circ[Th(SO_4)_2, c, 298\ K] = -551 \pm 6$ kcal/mol; $S^\circ[Th(SO_4)_2, c, 298\ K] = 40 \pm 5$ cal $\cdot$ mol^{-1} $\cdot$ K^{-1}.

The complete thermodynamic properties of $Th(SO_4)_2$ arising from the assessment made by Cordfunke and O'Hare [13] are given in Table 40, p. 65.

The following estimated values are given in the references cited: enthalpy of formation, -602 kcal/mol [14], entropy, 37.6 cal $\cdot$ mol^{-1} $\cdot$ K^{-1}, entropy of the crystal lattice, 130.6 cal $\cdot$ mol^{-1} $\cdot$ K^{-1} [15], heat capacity at 298 K, 41.3 cal $\cdot$ mol^{-1} $\cdot$ K^{-1}, heat capacity of the crystal lattice -6.5 cal $\cdot$ mol^{-1} $\cdot$ K^{-1} [16].

The estimated linear thermal expansion coefficient of $Th(SO_4)_2$ is 12.1×10^{-6} K^{-1} [16].

Infrared data for amorphous and crystalline $Th(SO_4)_2$ obtained by thermal decomposition of $Th(SO_3)_2 \cdot 4\ H_2O$ are given in [5]; the samples contained adsorbed water and no assignments were made (see also [6]).

Chemical Properties

The chemical reactions of thorium bis(sulfate) have barely been investigated.

It reacts with liquid NH_3 but the product was not studied prior to extraction with water, which yielded a solution of $(NH_4)_2SO_4$ and an unidentified residue [17]. The formation of hydrates is discussed below in Chapter 10.6.3 (p. 67) and the presently known sulfato complexes are discussed in Chapter 10.6.8 (p. 79).

$Th(SO_4)_2$ decomposes to give ThO_2 above ca. 750°C [1, 3, 5 to 7]. According to [4], the decomposition proceeds via the intermediate formation of $2\ ThO_2 \cdot 3\ SO_3$ ($\equiv Th_2O(SO_4)_3$) at 580°C; this observation has not been confirmed and no properties are given for the intermediate.

Thorium bis(sulfate) can be employed as a catalyst in the preparation of N-alkylated amines such as dodecylamine and dimethyldodecylamine [18 to 20], the isomerisation of cyclopropane [21], the formation of isoprene from the interaction of isobutylene and formaldehyde [22] and, as part of the mixture $SiO_2/Al_2O_3/BiPO_4/Th(SO_4)_2$, in the preparation of pyridine from formaldehyde, acetaldehyde, and ammonia [23].

References for 10.6.2:

[1] Mien-Ts'eng Su, Hsi Chün Hu (Beijing Daxue Xuebao Ziran Kexueban **3** [1957] 115/8; C.A. **1961** 17336). — [2] A. K. Molodkin, E. G. Arutyunyan (Zh. Neorgan. Khim. **10** [1965] 352/62; Russ. J. Inorg. Chem. **10** [1965] 189/95). — [3] A. K. Molodkin, G. A. Skotnikova, O. M. Ivanova (Zh. Neorgan. Khim. **11** [1966] 2241/54; Russ. J. Inorg. Chem. **11** [1966] 1201/9). — [4] V. T. Athavole, C. S. P. Iyer (J. Inorg. Nucl. Chem. **29** [1967] 1003/12) — [5] Yu. Ya. Kharitonov, V. N. Tverdokhlebov, A. K. Molodkin (Zh. Neorgan. Khim. **11** [1966] 2608/13; Russ. J. Inorg. Chem. **11** [1966] 1401/4).

[6] V. A. Golovnya, A. K. Molodkin, V. N. Tverdokhlebov (Zh. Neorgan. Khim. **12** [1967] 2377/8; Russ. J. Inorg. Chem. **12** [1967] 1254/9). — [7] V. A. Golovnya, A. K. Molodkin, V. N. Tverdokhlebov (Zh. Neorgan. Khim. **12** [1967] 2729/39; Russ. J. Inorg. Chem. **12** [1967] 1439/44). — [8] R. C. Paul, S. Singh, R. D. Verma (J. Indian Chem. Soc. **58** [1981] 24/5). — [9] S. W. Mayer, B. B. Owens, T. H. Rutherford, R. B. Serrins (NAA-SR-4783 [1960] 1/11; N.S.A. **14** [1960] No. 18934). — [10] G. Beck (Z. Anorg. Allgem. Chem. **174** [1928] 31/41).

[11] L. Wöhler, W. Plüddemann, P. Wöhler (Ber. Deut. Chem. Ges. **41** [1908] 703/17). — [12] M. Rand (in: O. Kubaschewski, Thorium, Physico-Chemical Properties of Its Compounds and Alloys, IAEA, Vienna 1975, pp. 7/85). — [13] E. H. P. Cordfunke, P. A. G. O'Hare (The Chemical Thermodynamics of Actinide Elements and Compounds. Pt. 3. Miscellaneous Actinide Compounds, IAEA, Vienna 1978). — [14] V. M. Amosov, V. E. Plyushchev (Izv. Vysshikh Uchebn. Zavadenii Khim. Khim. Tekhnol **11** [1968] 1128/34). — [15] A. I. Moskvin (Radiokhimiya **15** [1973] 353/62; Soviet Radiochem. **15** [1973] 356/63).

[16] A. I. Moskvin (Radiokhimiya **15** [1973] 362/6; Soviet Radiochem. **15** [1973] 364/7). — [17] G. W. Walt, W. A. Jenkins, J. M. McCuiston (J. Am. Chem. Soc. **72** [1950] 2260/2). — [18] S. Baron (U.S. 3732311 [1973] 1/2). — [19] S. Baron (F.R. 2168716 [1973] 1/5). — [20] S. Baron (U.S. 4076649 [1978] 1/3).

[21] Y. Imizu, H. Hattori, K. Tanabe, T. Kondo (Bull. Chem. Soc. Japan **52** [1979] 2189/91). — [22] T. Nakano, M. Uchida, F. Yoshitsugi (Japan. 7400162 [1964] 1/6; C.A. **81** [1974] No. 64950). — [23] Y. Wada, S. Yasuda, T. Niwa, Y. Tsuruta, T. Tagano (Japan. Kokai 76-63176 [1976] 1/6; C.A. **85** [1976] No. 192569).

10.6.3 Thorium Bis(sulfate) Hydrates, Th(SO$_4$)$_2$ · n H$_2$O (n = 9, 8, 6, 4.5?, 4, 2, and 1)

Several thorium bis(sulfate) hydrates, Th(SO$_4$)$_2$ · n H$_2$O (n = 9, 8, 6, (4.5?) 4, 2, and 1) are described in the literature. Those most readily obtained by crystallisation from aqueous media are Th(SO$_4$)$_2$ · 9 H$_2$O, Th(SO$_4$)$_2$ · 8 H$_2$O and Th(SO$_4$)$_2$ · 4 H$_2$O. The pre-1950 accounts of the different conditions (e.g. temperature, acidity, etc.) under which these phases have been prepared are discussed in "Thorium" 1955, pp. 286/89 (see also [1, 2]) together with the more limited information on the hexa- and dihydrate. The present discussion is confined essentially to the more recent publications and, for convenience, the different hydrates are dealt with in a single section.

Preparation

Th(SO$_4$)$_2$ · 9 H$_2$O crystallises at 0°C from a supersaturated solution obtained by dissolution of thorium hydroxide in dilute H$_2$SO$_4$ at 40 to 50°C; the rate of crystallisation can be increased and the solubility lowered by addition of 5 to 15 vol% of ethanol [3]. The nonahydrate has also been prepared by evaporation of slightly acid saturated solutions in the temperature range 15 to 25°C [4], and by addition of a 1.2 M H$_2$SO$_4$/0.75 M HNO$_3$ mixture to Th(NO$_3$)$_4$ in TBP-Varsol (TBP, tributylphosphate) [5]. In the latter instance the resulting precipitate was washed with a 1 M H$_2$SO$_4$/1 M HNO$_3$ mixture at ice temperature and air-dried. Similar procedures are described in references [6 to 8] which are concerned primarily with the separation of thorium from uranium. An alternative route to Th(SO$_4$)$_2$ · 9 H$_2$O involves crystallisation from thorium nitrate solutions to which the calculated quantity of H$_2$SO$_4$ was added; the acidity was maintained ⩽3.5 M [9].

Exposure of Th(SO$_4$)$_2$ · 4 H$_2$O to air results in conversion to Th(SO$_4$)$_2$ · 8 H$_2$O [4]. According to [10], the octahydrate is also obtained from solutions initially 1 M in H$_2$SO$_4$ by addition of sufficient ethanol to give a water:ethanol ratio of 1:2.

The tetrahydrate, Th(SO$_4$)$_2$ · 4 H$_2$O, has been obtained by allowing an 8.5 M H$_2$SO$_4$ solution of the nonahydrate to stand at room temperature for several days, and by addition of 3 M H$_2$SO$_4$ to a concentrated thorium nitrate solution at 60 to 80.°C [4], when crystallisation occurred within a few hours. The product was washed with hot water and rapidly dried to prevent hydration to the octahydrate. A product of composition Th(SO$_4$)$_2$ · 3.5 H$_2$O was obtained by washing the precipitate formed at room temperature from a sulfuric acid/thorium nitrate mixture with water, ethanol and ether [9].

The interaction of ThC or ThC_2 with 6 and 12 M H_2SO_4 at 80°C has given products in which the degree of hydration is reported to vary between 2.5 H_2O and 3.5 H_2O [11]. Other information on the precipitation of thorium bis(sulfate) hydrates of unspecified or unusual degree of hydration is available in references [12 to 14].

DTA and TGA studies involving $Th(SO_4)_2 \cdot 9\,H_2O$ [9, 15] and $Th(SO_4)_2 \cdot 7.2\,H_2O$ [9] (obtained by drying the nonahydrate to constant weight in air) have indicated the existence of lower hydrates. However, the mass-loss curves do not show good plateaux corresponding to the individual phases, indicating that the thermal stabilities of adjacent hydrates are close and that the lower hydrates are difficult to obtain pure by this technique [9].

Physical Properties

The only structural data available for thorium bis(sulfate) hydrates are those reported for $Th(SO_4)_2 \cdot 8\,H_2O$ [42], which crystallises in the monoclinic space group $P2_1/n\text{-}C_{2h}^5$ (No. 14) with   a = 8.51(2),   b = 11.86(2),   c = 13.46(2) Å   and   β = 92.65(1)°;   Z = 4, D_{calc} = 2.778 g/cm^3 and D_{meas} = 2.78 g/cm^3. The Th atom is 10-co-ordinated with both sulfate groups chelated and six of the water molecules being bonded to it. The thorium co-ordination polyhedron approximates closely to the bicapped square antiprism of D_{4d} symmetry and with θ = 63°. The environment around the thorium atom is illustrated in **Fig. 24**. The Th–O (sulfate) distances range from 2.53(2) to 2.58(2) Å and Th–O (water) bond lengths lie between 2.46(2) and 2.54(2) Å. The water molecules represented by O(1) and O(2) (not bonded to Th) and O(4), O(8), O(11), O(12), O(14), and O(15) (see Fig. 24) form hydrogen bonds among themselves and with the oxygens of the sulfate groups. The hydrogen bond distances range from 2.66(3) to 2.88(4) Å [42]. $Th(SO_4)_2 \cdot 8\,H_2O$ undergoes a reversible phase change at a pressure of 9.8 kbar at 25°C [16].

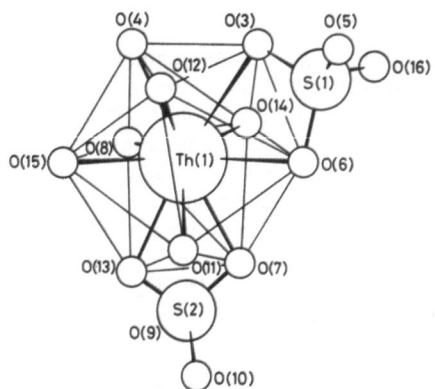

Fig. 24. Environment of the thorium atom in $Th(SO_4)_2 \cdot 8\,H_2O$ [42].

There have been no new experimental thermodynamic data for thorium bis(sulfate) hydrates since those reported by Koppel [17] for the enthalpies of solution of $Th(SO_4)_2 \cdot 8\,H_2O$ and $Th(SO_4)_2 \cdot 4\,H_2O$ in aqueous K_2CO_3 solution, −28.4 and −14.9 kcal/mol, respectively, and

Table 41
Infrared Data (in cm^{-1}) for Th(SO$_4$)$_2$ · 4 H$_2$O [23] and Th(SO$_4$)$_2$ · 8 H$_2$O [24].

Th(SO$_4$)$_2$ · 4 H$_2$O	Th(SO$_4$)$_2$ · 8 H$_2$O	assignment
400 w 420 vw	430	$\nu_2(\mathrm{SO_4^{2-}})$
590 m 605 m 615 m 642 m	555 603 634	$\nu_4(\mathrm{SO_4^{2-}})$
970 w, sh	980	$\nu_1(\mathrm{SO_4^{2-}})$
1040 sh 1100 s 1135 s 1227 m 1265 m	1035 1160 1200	$\nu_3(\mathrm{SO_4^{2-}})$
1635 m	1630 1663	δ (H$_2$O)
3300 s, b	3300 b	ν(OH)

s, strong; m, medium; w, weak; vw, very weak; sh, shoulder; b, broad.

for the water vapour pressures in the equilibrium $^1/_4$Th(SO$_4$)$_2$ · 8 H$_2$O (c) → $^1/_4$Th(SO$_4$)$_2$· 4 H$_2$O (c) + H$_2$O (g) which were shown to be 0.33 atm at 313.2 K and 0.080 atm at 325.2 K. Following the recent evaluations of thermodynamic data on thorium sulfate hydrates [18, 19], $\Delta H_{298}^0 = -13.52$ kcal/mol is the accepted value for the reaction Th(SO$_4$)$_2$ · 4 H$_2$O (c) + 4 H$_2$O (l) → Th(SO$_4$)$_2$ · 8 H$_2$O (c).

The following entropies (in cal · mol^{-1} · K^{-1}) [20] and heat capacities (in cal · mol^{-1} · K^{-1}) [21], at 298 K, have been estimated: 112.8 and 117.3, respectively, for Th(SO$_4$)$_2$ · 8 H$_2$O; 75.2 and 79.3, respectively, for Th(SO$_4$)$_2$ · 4 H$_2$O.

Magnetic susceptibility measurements on thorium bis(sulfate) hydrates, Th(SO$_4$)$_2$ · n H$_2$O (n = 8, 7.75, 7 and 4) have confirmed the expected diamagnetism although the reproducibilities of the values (χ_{mol} = ca. 190 ± 10, 184 ± 4, 179, and 121 ± 5 × 10^{-6} for the values of n quoted) were not good [22].

Infrared spectral data for Th(SO$_4$)$_2$ · 8 H$_2$O [23] and Th(SO$_4$)$_2$ · 4 H$_2$O [24] are listed in Table 41 together with tentative assignments.

In the photoelectron (ESCA) spectrum of Th(SO$_4$)$_2$ · 9 H$_2$O there are weak satellites at binding energies of ca. 5.5 eV from the primary photolines. Thus, the binding energies of the $4f_{7/2}$ and $4f_{5/2}$ peaks are 338.11 and 347.53 eV, respectively, with satellite separations of 5.7 and 5.4 eV, respectively [25]. The latter separation is reported elsewhere as 6.0 eV [26]. The satellites are assigned to ligand-to-thorium 5f shake-up transitions.

Chemical Properties

The bis(sulfate) hydrates readily lose water when heated, and DTA and TGA studies have given the following results. According to [15] Th(SO$_4$)$_2$ · 9 H$_2$O decomposes via the following hydrates: Th(SO$_4$)$_2$ · n H$_2$O, n = 4.5 at 80°C, 4 at 110°C, 2 at 180°C and 1 at 235°C. Very similar

References for 10.6.3 on pp. 74/5

results are reported in [9], except that $Th(SO_4)_2 \cdot 4.5 H_2O$ was not observed, viz: n = 4 at 75 to 120°C, 2 at 120 to 170°C, and 1 at 170 to 220°C. Decomposition of $Th(SO_4)_2 \cdot 7.25 H_2O$ (obtained by drying the nonahydrate to constant weight in air) proceeds via n = 4 at 60 to 130°C, n = 2 at 130 to 180°C and n = 1 at 180 to 220°C [9]. However, a commercial sample of composition $Th(SO_4)_2 \cdot 7.5 H_2O$ differed appreciably in behaviour, dehydration occurring in a single stage with an endothermic effect at 90 to 135°C [9]. The dehydration of a sample of composition $Th(SO_4)_2 \cdot 3.5 H_2O$ is reported to indicate the possible existence of hydrates intermediate between n = 1 and 0 but the deductions require confirmation [9].

Clearly defined plateaux corresponding to individual phases are not seen on the mass-loss versus temperature curves and it is obviously difficult to obtain any of the above-mentioned lower hydrates pure by thermal decomposition methods.

Earlier work on the dehydration of thorium bis(sulfate) hydrates is presented in "Thorium" 1955, pp. 286/9.

Solubility data reported prior to 1950 are given in "Thorium" 1955, pp. 286/9. The only recent publications appear to be those dealing with the phase systems $Th(SO_4)_2$-Na_2SO_4-H_2O [4] and $Th(SO_4)_2$-$Ce(SO_4)_2$-H_2O [27]. The former was studied at 25°C, at which temperature the hydrate observed was $Th(SO_4)_2 \cdot 9 H_2O$, and at 50°C, when the hydrate phase is $Th(SO_4)_2 \cdot 4 H_2O$. The results are illustrated in **Fig. 25** and **Fig. 26** and appropriate data are listed in Tables 42 and 43, p. 72.

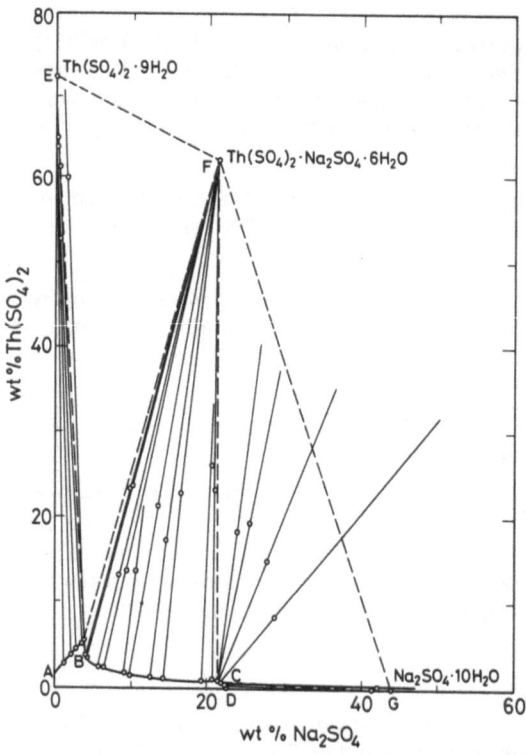

Fig. 25. 25° isotherm for the $Th(SO_4)_2$-Na_2SO_4-H_2O ternary system [4].

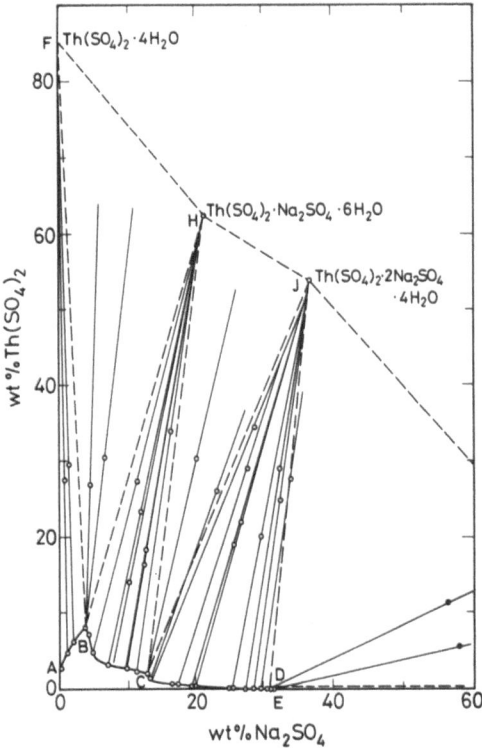

Fig. 26. 50° isotherm for the Th(SO$_4$)$_2$-Na$_2$SO$_4$-H$_2$O ternary system [4].

Table 42
The Thorium Sulfate-Sodium Sulfate-Water System at 25 °C [4].

liquid phase composition, wt %		"solid residue" composition, wt %		composition of solid phase
Th(SO$_4$)$_2$	Na$_2$SO$_4$	Th(SO$_4$)$_2$	Na$_2$SO$_4$	
1.63	—	—	—	Th(SO$_4$)$_2$ · 9 H$_2$O
2.6	1.2	64.9	0.2	Th(SO$_4$)$_2$ · 9 H$_2$O
3.6	2.1	64.0	0.2	Th(SO$_4$)$_2$ · 9 H$_2$O
4.3	2.9	52.9	0.6	Th(SO$_4$)$_2$ · 9 H$_2$O
5.0	3.7	61.7	0.7	Th(SO$_4$)$_2$ · 9 H$_2$O
5.3	4.0	60.2	1.5	Th(SO$_4$)$_2$ · 9 H$_2$O + Th(SO$_4$)$_2$ · Na$_2$SO$_4$ · 6 H$_2$O
3.6	4.1	23.7	10.3	Th(SO$_4$)$_2$ · Na$_2$SO$_4$ · 6 H$_2$O
3.2	4.3	23.5	9.9	Th(SO$_4$)$_2$ · Na$_2$SO$_4$ · 6 H$_2$O
2.3	5.9	13.3	8.2	Th(SO$_4$)$_2$ · Na$_2$SO$_4$ · 6 H$_2$O
2.3	6.6	13.9	9.5	Th(SO$_4$)$_2$ · Na$_2$SO$_4$ · 6 H$_2$O
1.7	9.2	13.8	10.5	Th(SO$_4$)$_2$ · Na$_2$SO$_4$ · 6 H$_2$O
1.4	9.8	21.4	13.5	Th(SO$_4$)$_2$ · Na$_2$SO$_4$ · 6 H$_2$O
1.2	12.4	17.5	14.6	Th(SO$_4$)$_2$ · Na$_2$SO$_4$ · 6 H$_2$O

Table 42 (continued)

liquid phase composition, wt%		"solid residue" composition, wt%		composition of solid phase
$Th(SO_4)_2$	Na_2SO_4	$Th(SO_4)_2$	Na_2SO_4	
1.1	14.1	23.0	16.6	$Th(SO_4)_2 \cdot Na_2SO_4 \cdot 6\,H_2O$
0.9	19.2	26.3	20.6	$Th(SO_4)_2 \cdot Na_2SO_4 \cdot 6\,H_2O$
1.0	19.5	—	—	$Th(SO_4)_2 \cdot Na_2SO_4 \cdot 6\,H_2O$
1.0	20.6	23.5	21.0	$Th(SO_4)_2 \cdot Na_2SO_4 \cdot 6\,H_2O$
0.8	21.2	18.5	23.8	$Th(SO_4)_2 \cdot Na_2SO_4 \cdot 6\,H_2O + Na_2SO_4 \cdot 10\,H_2O$
0.8	21.3	19.7	25.4	$Th(SO_4)_2 \cdot Na_2SO_4 \cdot 6\,H_2O + Na_2SO_4 \cdot 10\,H_2O$
0.9	21.3	15.2	27.7	$Th(SO_4)_2 \cdot Na_2SO_4 \cdot 6\,H_2O + Na_2SO_4 \cdot 10\,H_2O$
0.8	21.6	8.6	28.6	$Th(SO_4)_2 \cdot Na_2SO_4 \cdot 6\,H_2O + Na_2SO_4 \cdot 10\,H_2O$
0.7	21.7	0.3	42.9	$Na_2SO_4 \cdot 10\,H_2O$
0.4	21.9	0.01	42.1	$Na_2SO_4 \cdot 10\,H_2O$
—	22.9	—	—	$Na_2SO_4 \cdot 10\,H_2O$

Table 43
The Thorium Sulfate-Sodium Sulfate-Water System at 50°C [4].

liquid phase composition, wt%		"solid residue" composition, wt%		composition of solid phase
$Th(SO_4)_2$	Na_2SO_4	$Th(SO_4)_2$	Na_2SO_4	
2.5	—	—	—	$Th(SO_4)_2 \cdot 4\,H_2O$
4.7	1.0	27.5	0.7	$Th(SO_4)_2 \cdot 4\,H_2O$
6.1	1.9	29.7	1.4	$Th(SO_4)_2 \cdot 4\,H_2O$
8.0	3.6	27.0	4.4	$Th(SO_4)_2 \cdot 4\,H_2O + Th(SO_4)_2 \cdot Na_2SO_4 \cdot 6\,H_2O$
8.0	3.6	30.5	6.5	$Th(SO_4)_2 \cdot 4\,H_2O + Th(SO_4)_2 \cdot Na_2SO_4 \cdot 6\,H_2O$
7.1	4.1	32.3	9.3	$Th(SO_4)_2 \cdot Na_2SO_4 \cdot 6\,H_2O$
4.8	4.7	27.5	11.2	$Th(SO_4)_2 \cdot Na_2SO_4 \cdot 6\,H_2O$
3.1	7.1	23.7	11.5	$Th(SO_4)_2 \cdot Na_2SO_4 \cdot 6\,H_2O$
3.3	7.7	14.1	10.1	$Th(SO_4)_2 \cdot Na_2SO_4 \cdot 6\,H_2O$
3.0	9.5	18.3	12.7	$Th(SO_4)_2 \cdot Na_2SO_4 \cdot 6\,H_2O$
2.5	9.6	16.5	12.3	$Th(SO_4)_2 \cdot Na_2SO_4 \cdot 6\,H_2O$
2.3	11.1	34.0	16.1	$Th(SO_4)_2 \cdot Na_2SO_4 \cdot 6\,H_2O$
2.0	12.7	30.5	20.0	$Th(SO_4)_2 \cdot Na_2SO_4 \cdot 6\,H_2O + Th(SO_4)_2 \cdot 2\,Na_2SO_4 \cdot 4\,H_2O$
1.4	13.2	26.2	22.8	$Th(SO_4)_2 \cdot 2\,Na_2SO_4 \cdot 4\,H_2O$
1.2	13.3	34.9	28.4	$Th(SO_4)_2 \cdot 2\,Na_2SO_4 \cdot 4\,H_2O$
0.4	16.2	17.9	23.4	$Th(SO_4)_2 \cdot 2\,Na_2SO_4 \cdot 4\,H_2O$
0.5	17.3	29.6	27.3	$Th(SO_4)_2 \cdot 2\,Na_2SO_4 \cdot 4\,H_2O$
0.3	19.2	19.3	25.3	$Th(SO_4)_2 \cdot 2\,Na_2SO_4 \cdot 4\,H_2O$
0.2	19.5	22.2	26.3	$Th(SO_4)_2 \cdot 2\,Na_2SO_4 \cdot 4\,H_2O$
<0.01	24.9	38.8	34.5	$Th(SO_4)_2 \cdot 2\,Na_2SO_4 \cdot 4\,H_2O$
<0.01	25.1	20.5	29.3	$Th(SO_4)_2 \cdot 2\,Na_2SO_4 \cdot 4\,H_2O$

Table 43 (continued)

liquid phase composition, wt%		"solid residue" composition, wt%		composition of solid phase
Th(SO$_4$)$_2$	Na$_2$SO$_4$	Th(SO$_4$)$_2$	Na$_2$SO$_4$	
<0.01	27.1	29.9	32.0	Th(SO$_4$)$_2$ · 2 Na$_2$SO$_4$ · 4 H$_2$O
<0.01	28.3	25.2	32.1	Th(SO$_4$)$_2$ · 2 Na$_2$SO$_4$ · 4 H$_2$O
0.02	29.3	28.2	33.9	Th(SO$_4$)$_2$ · 2 Na$_2$SO$_4$ · 4 H$_2$O
0.1	30.2	6.0	58.3	Th(SO$_4$)$_2$ · 2 Na$_2$SO$_4$ · 4 H$_2$O + Na$_2$SO$_4$
0.1	30.8	11.8	56.3	Th(SO$_4$)$_2$ · 2 Na$_2$SO$_4$ · 4 H$_2$O + Na$_2$SO$_4$
0.2	30.7	1.5	72.0	Na$_2$SO$_4$
0.2	30.8			Na$_2$SO$_4$
—	31.9			Na$_2$SO$_4$

The Th(SO$_4$)$_2$-Ce(SO$_4$)$_2$-H$_2$O phase system [27], in which no complex formation occurs, is shown in **Fig. 27**.

The interaction of solid Th(SO$_4$)$_2$ · 4 H$_2$O with aqueous NH$_3$ or aqueous NaOH is reported to result in the formation of hydroxide oxide species still containing small amounts of sulfate [28] (see also [29]).

Th(SO$_4$)$_2$ · 8 H$_2$O reacts with dimethylsulfoxide and urea at room temperature in the absence of a solvent to give complexes such as Th(SO$_4$)$_2$ · 4 (CH$_3$)$_2$SO · 3 H$_2$O [30, 31], Th(SO$_4$)$_2$ · 4 CO(NH$_2$)$_2$ · H$_2$O, Th(SO$_4$)$_2$ · 5 CO(NH$_2$)$_2$ · 3 H$_2$O, and Th(SO$_4$)$_2$ · 8 CO(NH$_2$)$_2$ · x H$_2$O [24,32] (see "Thorium" Suppl. Vol. E, 1985, pp. 103, 38). Reactions with the latter ligand at 130 to 135°C give anhydrous 1:4, 1:5, and 1:8 complexes [32]. The addition of hydrazine hydrate to a suspension of "Th(SO$_4$)$_2$ · 2 H$_2$O" in CCl$_4$ gives Th(SO$_4$)$_2$ · 2 N$_2$H$_4$ (see "Thorium" Suppl. Vol. E, 1985, p. 9) [33].

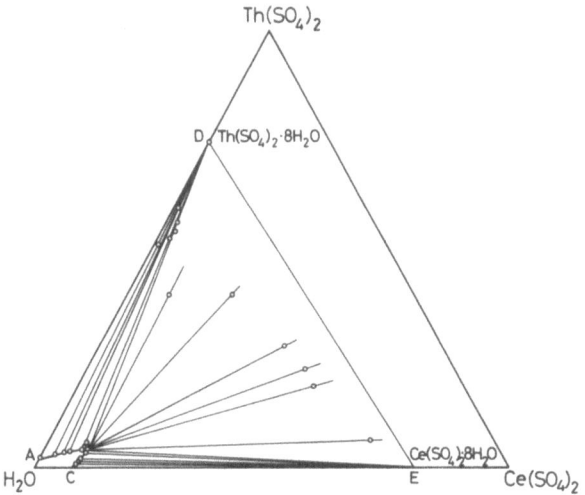

Fig. 27. The Th(SO$_4$)$_2$-Ce(SO$_4$)$_2$-H$_2$O system at 25°C [27].

The preparation of sulfatothorates(IV) is discussed in Chapter 10.6.8 (p. 79).

Th^{IV} in aqueous solution, prepared by dissolution of $Th(SO_4)_2 \cdot 4\,H_2O$, is reported to have been reduced by pulse radiolysis through a one electron reaction according to $ThSO_4^{2+}$ + $e_{aq}^- \rightarrow ThSO_4^+$. The rate constant at zero ionic strength is $1.0 \pm 0.5 \times 10^{10}\,mol^{-1} \cdot s^{-1}$ [43].

The use of thorium bis(sulfate) hydrates to catalyse nitrate formation by Nitrosomonas [34] and Nitrobacter agilis [35] and the hydrolysis of 5′-ribonucleotides to ribonucleosides [36] has been studied. It has also been shown that rolled oats to which thorium bis(sulfate) hydrates mixed with 0.1% dehydroacetic acid (= 3-acetyl-6-methyl-pyran-2,4-dion) were added were effective in rat control in sewers [37]. The effects of thorium sulfate in aqueous solution on coagulation of sols [38, 39], the agglomeration of crystalline ammonium and potassium sulfate [40], and the miscibility of phenol and water [41] have also been studied.

References for 10.6.3:

[1] J. W. Mellor (A Comprehensive Treatise on Inorganic and Theoretical Chemistry, Vol. VII, Longmans, London 1947, pp. 240/8). − [2] J. Flahaut (in: P. Pascal, Nouveau Traité de Chimie Minérale, Vol. IX, Paris 1967, pp. 1103/9). − [3] P. Krumolz, F. Gottdenker (Proc. Intern. Conf. Peaceful Uses At. Energy, Geneva 1955, Vol. 8, pp. 126/8). − [4] Huang-Pang Chang, P. I. Fedorov (Zh. Neorgan. Khim. **6** [1961] 971/6; Russ. J. Inorg. Chem. **6** [1961] 494/7). − [5] K. J. Bril, P. G. De Saboia Araujo (Ind. Eng. Chem. Process Design Develop. **3** [1964] 8/10).

[6] K. J. Bril, P. Krumolz (Brit. 880046 [1961]). − [7] K. J. Bril, P. Krumolz (Ger. 1163305 [1964] 1/6). − [8] K. J. Bril, P. Krumolz (U.S. 3104940 [1961]). − [9] A. K. Molodkin, O. M. Ivanova (Zh. Neorgan. Khim. **11** [1966] 2241/54; Russ. J. Inorg. Chem. **11** [1966] 1201/9). − [10] B. D. Stepin, G. M. Gulyaev, A. I. Chernyak (Khim. Nauka Prom. **4** [1959] 681/2; C.A. **1960** 7394).

[11] M. B. Sears, L. M. Ferris (J. Inorg. Nucl. Chem. **29** [1967] 1548/52) − [12] Societé de Produits Chimiques des Terres Rares (Brit. 674400 [1952]; C.A. **1952** 11610). − [13] C. W. Kline, W. R. Bennett (U.S. 3047359 [1962] 1/4). − [14] J. C. Geertsma, R. V. Pammenter (S. African 6903756 [1970] 1/8). − [15] Mien-Ts'eng Su, Hsi Chün Hu (Beijing Daxue Xuebao Ziran Kexueban **3** [1957] 115/8).

[16] R. R. Sood, R. A. Stager (Science **154** [1966] 388/90). − [17] I. Koppel (Z. Anorg. Allgem. Chem. **67** [1910] 293/301; AEC-tr-5385 [1962] 1/13; N.S.A. **17** [1963] No. 152). − [18] M. Rand (in: O. Kubaschewski, Thorium, Physico-Chemical Properties of Its Compounds and Alloys, IAEA, Vienna 1975, pp. 7/85). − [19] E. H. P. Cordfunke, P. A. G. O'Haire (The Chemical Thermodynamics of Actinide Elements and Compounds, Pt. 3, Miscellaneous Actinide Compounds, IAEA, Vienna 1978). − [20] A. I. Moskvin (Radiokhimiya **15** [1973] 353/62: Soviet Radiochem. **15** [1973] 356/63).

[21] A. I. Moskvin (Radiokhimiya **15** [1973] 362/6; Soviet Radiochem. **15** [1973] 364/7). − [22] V. I. Belova, Ya. K. Syrkin, A. K. Molodkin, O. M. Ivanova, L. M. Shiporina (Zh. Neorgan. Khim. **13** [1968] 1458/60; Russ. J. Inorg. Chem. **13** [1968] 766/7). − [23] O. N. Evstaf'eva, A. K. Molodkin, G. G. Dvoryantseva, O. M. Ivanova, M. I. Struchkova (Zh. Neorgan. Khim. **11** [1966] 1306/15; Russ. J. Inorg. Chem. **11** [1966] 697/702). − [24] K. I. Petrov, A. K. Molodkin, O. D. Saralidze, O. M. Ivanova (Zh. Neorgan. Khim. **14** [1969] 1227/31; Russ. J. Inorg. Chem. **14** [1969] 643/5). − [25] G. M. Bancroft, T. K. Sharma, J. L. Esquivel, S. Larsson (Chem. Phys. Letters **51** [1977] 105/10).

[26] G. C. Allen, P. M. Tucker (Chem. Phys. Letters **43** [1976] 254/7). − [27] Wei-Pong Tsung, Yee-Chen Sun (Huaxue Xuebao **24** [1958] 274/6). − [28] V. V. Sakharov, T. I. Danilevich, V. M. Klyuchnikov, G. N. Voronskaya, S. S. Korovin (Radiokhimiya **16** [1974] 74/81; Soviet Radiochem. **16** [1974] 70/5). − [29] R. P. Singh, N. R. Banerjee (J. Indian Chem. Soc. **39** [1962]

255/9). — [30] A. K. Molodkin, O. M. Ivanova, Z. V. Belyakova, L. E. Kolesnikova (Zh. Neorgan. Khim. **15** [1970] 3245/6; Russ. J. Inorg. Chem. **15** [1970] 1692/3).

[31] K. I. Petrov, O. M. Ivanova, A. K. Molodkin (Zh. Neorgan. Khim. **17** [1972] 1613/5; Russ. J. Inorg. Chem. **17** [1972] 834/6). — [32] A. K. Molodkin, O. M. Ivanova, L. E. Kozina (Zh. Neorgan. Khim. **13** [1968] 2308/9; Russ. J. Inorg. Chem. **13** [1968] 1192/3). — [33] V. T. Athavale, C. S. P. Iyer (J. Inorg. Nucl. Chem. **29** [1967] 1003/12). — [34] S. P. Tandon, S. K. De, R. C. Rastogi (Agrokem. Talajtan **12** [1963] 293/8; C.A. **60** [1964] 5914). — [35] S. P. Tandon, M. M. Mishra (Zentralbl. Bakteriol. Parasitenk. Infektionskr. Hyg. II **122** [1968] 155/7; C.A. **69** [1968] No. 41960).

[36] H. Mori, S. Kusui, A. Murakoshi (Japan. Kokai 7553378 [1975] 1/8; C.A. **83** [1975] No. 114834). — [37] K. Becker (Z. Angew. Zool. **52** [1965] 173/96). — [38] R. Kumar, A. K. Bhattacharya (J. Indian Chem. Soc. **28** [1951] 638/44). — [39] K. F. Schulz, E. Matjevic (Kolloid Z. **168** [1960] 143/50). — [40] E. G. Cooke, L. Phoenix (Brit. 917567 [1963] 1/6).

[41] S. T. Bowden, J. H. Purnell (J. Chem. Soc. **1954** 535/8). — [42] J. Habash, A. J. Smith (Acta Cryst. C **39** [1983] 413/5). — [43] D. Martin-Rovet, A. M. Koulkess-Pujo, M. Plissonnier, G. Folcher (Radiat. Phys. Chem. **21** [1983] 473/9).

10.6.4 Thorium Oxide Sulfate, ThOSO$_4$

The white oxide sulfate, ThOSO$_4$, has recently been reported to form on the thermal decomposition of Th(O$_2$)SO$_4$ · 3 H$_2$O (see p. 76) at 500°C [1, 2]. X-ray powder diffraction data are listed in [2]; these have not been interpreted. Infrared spectral results (see Table 44) are claimed to be consistent with a dimeric unit containing bridging oxygen atoms and bidentate sulfate groups [1, 2]. In view of the resulting low coordination number, 4, for the thorium atom these proposals should be viewed with caution.

ThOSO$_4$ decomposes to ThO$_2$ above 690°C [1, 2].

Table 44
Infrared Data for ThOSO$_4$ and Th(O$_2$)SO$_4$ · 3 H$_2$O, in cm^{-1} [1].

ThOSO$_4$	Th(O$_2$)SO$_4$ · 3 H$_2$O	assignment
—	3480 s, b	ν(O-H) of lattice water
—	3240 s, vb	ν(O-H) of co-ordinated water
—	1680 s	δ(H-O-H) of co-ordinated water
—	1620 s	δ(H-O-H) of lattice water
1200 s	1200 s	
1145 s	1150 s	ν$_3$(SO$_4$)
1085 s	1085 s	
1020 s	970 s	ν$_1$(SO$_4$)
770 w	—	
725 m	—	M-O stretch of dioxo-bridged structure
675 m	—	
655 w	645 w	
605 s	620 s	ν$_4$(SO$_4$)
590 w	590 w	
450 w	450 w	ν$_2$(SO$_4$)

References for 10.6.4 on p. 76

There have been no further publications concerning the oxide sulfate hydrates, $ThOSO_4 \cdot$ $n\,H_2O$ ($n = 5, 3, 2, 1$), since they were discussed in "Thorium" 1955, pp. 292/3, and there remains the possibility that these were hydroxide rather than oxide sulfates (cf. $Th(OH)_2SO_4$, Chapter 10.6.6).

References for 10.6.4:

[1] V. Raman, G. V. Jere (Indian J. Chem. **11** [1973] 31/4). — [2] V. Raman, G. V. Jere (Indian J. Chem. **11** [1973] 476/7).

10.6.5 Thorium Peroxide Sulfate Hydrates

The only well characterised thorium peroxide sulfate is the trihydrate, $Th(O_2)SO_4 \cdot 3\,H_2O$, which has been obtained by precipitation from nitrate sulfate or sulfate media initially >0.2 M in acid [1 to 3].

No structural data are available for this compound. Infrared spectral results (see Table 44, p. 75) have been interpreted on the basis of bridging peroxo groups and bidentate sulfate groups. Raman data are not available to confirm the former assignment, which is based on the absence of an IR mode at ca. 880 cm^{-1}. A coordination number of 6 is suggested for the trihydrate on the basis of the IR data and thermal decomposition studies which are claimed to show that two of the three water molecules are co-ordinated to the thorium atom [3]. An X-ray structure determination is required to confirm this low co-ordination number.

The oxide sulfate, $ThOSO_4$, is formed when $Th(O_2)SO_4 \cdot 3\,H_2O$ is heated at 500°C [3, 4]; at higher temperatures, >690°C, ThO_2 is obtained.

A range of phases containing differing amounts of peroxide, nitrate, sulfate, and water have been precipitated from low acid media [2] but apart from analysis and X-ray studies, which showed them to be isomorphous with thorium peroxide nitrates, they have not been studied.

References for 10.6.5:

[1] J. W. Hamaker, C. W. Koch (TID-5223 [1952] 318/38; N.S.A. **11** [1957] No. 11493). — [2] J. W. Hamaker, C. W. Koch (TID-5223 [1952] 339/47; N.S.A. **11** [1957] No. 11493). — [3] V. Raman, G. V. Jere (Indian J. Chem. **11** [1973] 31/4). — [4] V. Raman, G. V. Jere (Indian J. Chem. **11** [1973] 476/7).

10.6.6 Thorium Hydroxide Sulfates

Thorium dihydroxide sulfate, $Th(OH)_2SO_4$, is prepared by heating an aqueous solution of the bis(sulfate) in a sealed tube at 105 to 115°C [1]. The resulting white, crystalline compound possesses orthorhombic symmetry, space group Pnma-D_{2h}^{16} (No. 62) with a = 11.733(5), b = 6.040(5), and c = 7.059(5) Å. The structure consists of zigzag chains of $Th(OH)_2$ groups held together by the sulfate groups. The thorium atoms are 8-coordinate, with square antiprismatic stereochemistry involving four bonds to hydroxide groups and four to sulfates [1].

Aging of thorium nitrate solutions containing sodium sulfate (1:1 mole ratio Th:SO_4) and of thorium sulfate solutions, at elevated temperature within an initial pH range of 3.8 to 2.5, has yielded spherical colloidal particles of a composition close to $Th(OH)_2SO_4 \cdot H_2O$ [2]. For equivalent periods of aging, removal of thorium from solutions was greatest at the highest pH value. The analytical data indicate partial replacement of sulfate (in Th(OH)$_2$SO$_4$ · H$_2$O, for example) by hydroxide groups or that the product is a mixture of the basic sulfate and thorium hydroxide.

The earlier studies on basic sulfates are discussed in "Thorium" 1955, pp. 292/3.

References for 10.6.6:

[1] G. Lundgren (Arkiv Kemi **2** [1950] 535/49). — [2] N. B. Milic, E. Matijević (J. Colloid Interface Sci. **85** [1982] 306/15).

10.6.7 Thorium(IV) Mixed Acid Sulfates

10.6.7.1 Thorium(IV) Oxydiacetate Sulfate Hydrate, Th{O(CH$_2$COO)$_2$}SO$_4$ · 3H$_2$O

The addition of an excess of a sulfuric acid solution of Na$_2${O(CH$_2$COO)$_2$} to an aqueous solution of thorium nitrate gives the white solid Th{O(CH$_2$COO)$_2$}SO$_4$ · 3H$_2$O, which possesses monoclinic symmetry, space group C2-C$_2^3$ (No. 5), with a = 10.67(1), b = 8.35(1), c = 6.73(1) Å, β = 110.96(2)°, Z = 2, D$_{calc}$ = 3.05, D$_{meas}$ = 3.05(1) g/cm^3. The thorium atom is nine-co-ordinate, the stereochemistry being a monocapped square antiprism. Each oxydiacetate is chelated to the same Th atom through the ether oxygen (Th-O, 2.63(1) Å) and two carboxylate oxygen atoms (Th-O, 2.41(1) Å). Two of the three water molecules are co-ordinated (Th-O, 2.44(1) Å) and the sulfate groups bridge thorium atoms in the direction of the c axis (Th-O, 2.38(1) Å). A projection of the compact structural array is shown in **Fig. 28**. Infrared spectra of the substance

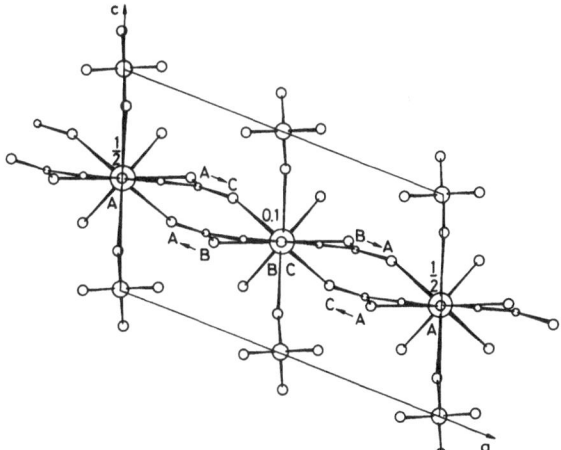

Fig. 28. Bonding scheme for Th{O(CH$_2$COO)$_2$}SO$_4$ · 3 H$_2$O; projection along the b axis. Thorium atoms at y = $^1/_2$ (A) are bridged to Th at y = 0 (B) and y = 1 (C) along the a axis. Thorium atoms B and C are superimposed in the figure. The oxydiacetate ligand of B bridges A and that of A bridges C; thus each Th is linked through the oxydiacetate ligands to four Th atoms at z = $^1/_2$ along a and through the sulfate groups to the two other Th atoms along the x axis of the cell [1].

                    References for 10.6.7 on p. 79

in CsI and polythene discs have been measured between 4000 to 180 and 300 to 50 cm^{-1}. A normal co-ordinate vibrational analysis on the Th$\{O(CH_2COO)_2\}$ entity, neglecting intermolecular effects except the carboxylate-thorium ones, has yielded the Urey-Bradley force constants listed in Table 45. The atom numbering scheme and the molecular model of the repeating unit used for the analysis are shown in **Fig. 29** [1].

Table 45
Urey-Bradley Force Constants in 10^2 N/m for Th^{IV} Oxydiacetate Sulfate Hydrate [1].

stretching			bending			repulsive		
K_1	Th-O(2)	1.63	H_1	Th-O(2)-C(1)	0.06	F_1	Th...C(1)	0.05
K_2	O(2)-C(1)	8.98	H_2	O(2)-Th-O(3)	0.07	F_2	O(2)...O(3)	0.08
K_3	C(1)-C(2)	2.77	H_3	O(2)-Th-O(2^{II})	0.08	F_3	O(2)...O(2^{II})	0.08
K_4	C(1)-O(1)	8.81	H_4	O(1)-C(1)-C(2)	0.28	F_4	O(1)...C(2)	0.54
K_5	C(2)-O(3)	3.98	H_5	O(1)-C(1)-O(2)	0.21	F_5	O(1)...O(2)	2.39
K_6	C(2)-H	4.02	H_6	O(2)-C(1)-C(2)	0.68	F_6	O(2)...C(2)	0.73
K_7	Th-O(3)	1.37	H_7	C(1)-C(2)-O(3)	0.85	F_7	C(1)...O(3)	0.15
K_8	Th-O(1^I)	1.30	H_8	C(1)-C(2)-H	0.30	F_8	C(1)...H	0.29
			H_9	C(1)-O(1)-Th^I	0.06	F_9	C(1)...Th	0.07
			H_{10}	H-C(2)-O(3)	0.22	F_{10}	H...O(3)	0.77
			H_{11}	C(2)-O(3)-Th	0.05	F_{11}	C(2)...Th	0.03
			H_{12}	C(2)-O(3)-C(2^{II})	0.31	F_{12}	C(2)...C(2^{II})	0.43
			H_{13}	H-C(2)-H	0.38	F_{13}	H...H	0.14

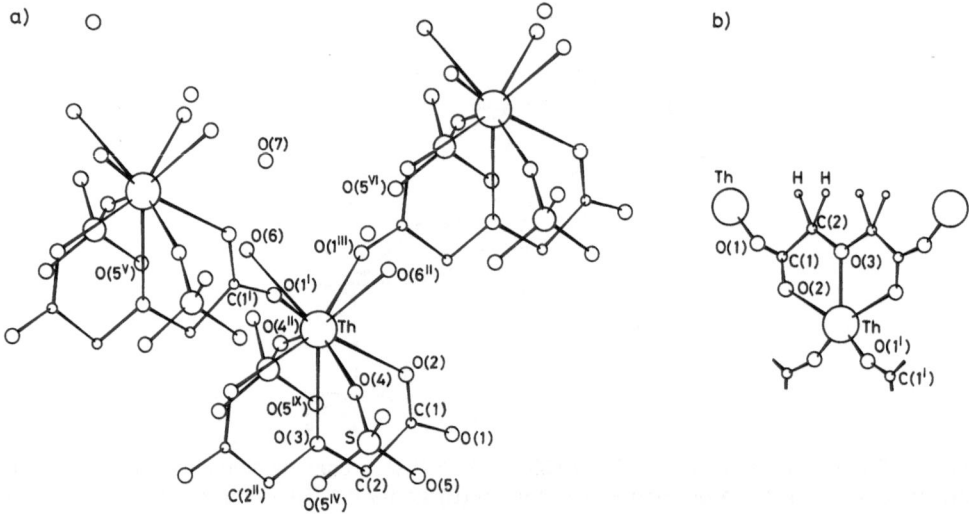

Fig. 29. (a) Crystal structure of [Th$\{O(CH_2COO)_2\}SO_4(H_2O)_2$] · H$_2$O showing the atomic numbering. (b) Molecular model of the repeating unit used for normal co-ordinate analysis [1].

Reference for 10.6.7:

R. Graziani, G. A. Battiston, U. Casellato, G. Sbrignadello (J. Chem. Soc. Dalton Trans. **1983** 1/7).

10.6.8 Thorium(IV) Sulfato Complexes

The following sulfato complexes have been reported: $M_2^I Th(SO_4)_3 \cdot n\, H_2O$ (M^I = Na with n = 12, 6, 4 and 0; K with n = 4 and 0; Rb with n = 2 and 0; Cs with n = 2 and 0; Tl with n = 4; NH_4 with n = 5, 4 and 0; $C_2H_5NH_3$ with n = 4; $(C_2H_5)_2NH_2$ with n = 4; C_5H_6N with n = 4; C_9H_8N with n = 4), $Na(CN_3H_6)Th(SO_4)_3 \cdot n\, H_2O$ (n = 6 and 0), $M_4^I Th(SO_4)_4 \cdot n\, H_2O$ (M^I = Na with n = 4; K with n = 2 and 0; Rb with n = 0; Cs with n = 1 and 0; NH_4 with n = 2 and 0; $C_6H_5N_2H_4$ (phenylhydrazinium) with n = 0; $C_8H_{17}NH_3$ with n = 0), $SnTh(SO_4)_4 \cdot 2\, H_2O$, $M_6^I Th(SO_4)_5 \cdot n\, H_2O$ (M^I = K with n = 3; Rb with n = 3 and 0; NH_4 with n = 3 and 0), $[Co(NH_3)_6]_2Th(SO_4)_5 \cdot 2\, H_2O$, $M_8^I Th(SO_4)_6 \cdot n\, H_2O$ (M^I = K with n = 2; NH_4 with n = 2 and 0), $Na_{10}Th_2(SO_4)_9 \cdot n\, H_2O$ (n = 5 and 0) and $M_{14}^I Th_2(SO_4)_{11}$ (M^I = K and Tl).

The pre-1950 literature is covered in "Thorium" 1955, pp. 325, 333/4, 338/9, 343, and 344/6, where details of the early preparations of many of the above compounds are given together with, where available, information on the solubilities in the $M_2^I SO_4$-$Th(SO_4)_2$-H_2O systems (see also [1, 2]). The present account covers only the subsequent publications and since several of the complexes have not been studied recently (see Table 39, p. 64) the reader is referred to "Thorium" 1955 for references to the original literature on such complexes.

Preparation

An investigation of the $Th(SO_4)_2 \cdot 9\, H_2O$-$Na_2SO_4$-$H_2O$ and $Th(SO_4)_2 \cdot 4\, H_2O$-$Na_2SO_4$-$H_2O$ systems at 25 and 50 °C, respectively, has shown that $Na_2Th(SO_4)_3 \cdot 6\, H_2O$ crystallises in both systems and that $Na_2Th(SO_4)_3 \cdot 4\, H_2O$ also crystallises at 50 °C [3]. The systems are illustrated in Fig. 25 and Fig. 26 (pp. 70 and 71), and details are also provided in Tables 42 and 43 (pp. 71 and 72). An alternative route to the hexahydrate involves addition of a solution of thorium tetranitrate to a slightly acid solution of sodium sulfate [4]. Other tris(sulfato) complexes obtained by the interaction of $Th(SO_4)_2 \cdot 8\, H_2O$ with aqueous solutions of univalent sulfates are: $(NH_4)_2Th(SO_4)_3 \cdot 5\, H_2O$, $K_2Th(SO_4)_3 \cdot 4\, H_2O$, $Rb_2Th(SO_4)_3$, and $Cs_2Th(SO_4)_3 \cdot 2\, H_2O$ [4]. In all instances the white products were air-dried. With a mixture of sodium sulfate and guanidinium sulfate the product is $(CN_3H_6)NaTh(SO_4)_3 \cdot 6\, H_2O$ [5].

Dehydration of the alkali metal complexes, $(NH_4)_2Th(SO_4)_3 \cdot 5\, H_2O$ and $(CN_3H_6)NaTh(SO_4)_3 \cdot 6\, H_2O$, is observed when they are heated at moderate temperatures. The anhydrous salts have been obtained during DTA and TGA studies at the temperatures shown in Table 46 (p. 80) [4, 5, 6]. In certain instances the water may be completely removed by longer periods of heating at lower temperatures than those quoted. Although the DTA curves show endotherms indicating stepwise removal of water from $Na_2Th(SO_4)_3 \cdot 6\, H_2O$ and $K_2Th(SO_4)_3 \cdot 4\, H_2O$ [4, 6] there are no plateaux in the TGA curves and intermediate hydrates have not been characterised during thermal dehydration studies.

References for 10.6.8 on pp. 86/7

Table 46

Refractive Indices, Densities and Dehydration Temperatures for Hydrated Sulfatothorates(IV).

	n_g	n_m	n_p	n_1	n_2	Ref.	D_{meas} in g/cm³	Ref.	T_A[1] in °C	Ref.
$Na_2Th(SO_4)_3 \cdot 6\ H_2O$				1.532	1.582	[4]	3.060	[4]	400	[6]
$K_2Th(SO_4)_3 \cdot 4\ H_2O$									235	[6]
$(NH_4)_2Th(SO_4)_3 \cdot 5\ H_2O$	1.592	1.557	1.541			[4]			265	[6]
$Rb_2Th(SO_4)_3 \cdot 2\ H_2O$									220	[6]
$Cs_2Th(SO_4)_3 \cdot 2\ H_2O$									295	[6]
$(CN_3H_6)NaTh(SO_4)_3 \cdot 6\ H_2O$							2.623	[5]	170	[5]
$K_4Th(SO_4)_4 \cdot 2\ H_2O$	1.557	1.545	1.540			[8]	3.27	[8, 13]	190	[6]
$(NH_4)_4Th(SO_4)_4 \cdot 2\ H_2O$	1.561	1.553	1.551			[8]	2.462	[8]	145	[6]
$Rb_4Th(SO_4)_4$				1.522	1.553	[8]	3.567	[8]	—	
$Cs_4Th(SO_4)_4 \cdot H_2O$							3.895	[8]	295	[6]
$Na_{10}Th_2(SO_4)_9 \cdot 5\ H_2O$	1.543	1.537	1.531			[8]	3.148	[8]	295	[6]
$(NH_4)_6Th(SO_4)_5 \cdot 3\ H_2O$	1.532	1.508	1.505			[9]	2.283	[9]	170	[6]
$Cs_6Th(SO_4)_5 \cdot 3\ H_2O$	1.538	1.528	1.519			[9]	3.675	[9]	180	[6]
$(NH_4)_8Th(SO_4)_6 \cdot 2\ H_2O$	1.538	1.533	1.525			[9]	2.267	[9]	180	[6]

1) T_A = temperature for conversion to anhydrous sulfato complex. The listed temperatures were recorded during DTA and TGA studies; dehydration can be often achieved at lower temperatures on prolonged heating.

Reactions involving 1:4 mole ratios of either thorium sulfate octahydrate or thorium nitrate solution and the appropriate alkali, etc. sulfate have been employed to prepare the tetrakis(sulfato) complexes $(NH_4)_4Th(SO_4)_4 \cdot 2\,H_2O$, $K_4Th(SO_4)_4 \cdot 2\,H_2O$, $Rb_4Th(SO_4)_4$, and $Cs_4Th(SO_4)_4 \cdot H_2O$ [8]. The potassium salt is readily dehydrated, being converted to $K_4Th(SO_4)_4 \cdot H_2O$ and $K_4Th(SO_4)_4$ when washed with ether or methanol, respectively [8].

The hydrated tetrakis(sulfato) complexes also lose water readily on being heated; the anhydrous salts are formed [6, 8] at the temperatures listed in Table 46. The TGA data show no evidence for the formation of intermediate hydrates.

Solvent extraction studies have indicated the formation of $(C_8H_{17}NH_3)_4Th(SO_4)_4$ in $CHCl_3$, but the solid complex does not appear to have been isolated [7].

The complex of unusual composition, $Na_{10}Th_2(SO_4)_9 \cdot 5\,H_2O$, is reported to crystallise from a slightly acidic solution of $Th(SO_4)_2 \cdot 8\,H_2O$ and Na_2SO_4 (1:6 mole ratio) [8]. Similar methods have been employed for the preparation of the pentakis- and hexakis(sulfato) complexes $(NH_4)_6Th(SO_4)_5 \cdot 3\,H_2O$ [9], $Cs_6Th(SO_4)_5 \cdot 3\,H_2O$ [9], $[Co(NH_3)_6]_2Th(SO_4)_5 \cdot 2\,H_2O$ [10, 11], and $(NH_4)_8Th(SO_4)_6 \cdot 2\,H_2O$ [9].

The formation of anhydrous complexes from the above hydrates has been observed, during DTA and TGA studies, at the temperatures shown in Table 46 [6, 8, 9]. $Cs_6Th(SO_4)_5 \cdot 3\,H_2O$ and $(NH_4)_8Th(SO_4)_6 \cdot 2\,H_2O$ are also completely dehydrated when stored in a desiccator over concentrated H_2SO_4 [9].

Attempts to isolate thorium(IV) sulfato complexes containing the divalent cations Cu[II], Fe[II], and Ni[II] were unsuccessful [9].

Physical Properties

Full structural data are available only for the complex $K_4Th(SO_4)_4 \cdot 2\,H_2O$ [12, 13]. This compound crystallises in the triclinic space group $P\bar{1}\text{-}C_i^1$ (No. 2) with a = 10.096, b = 16.752, c = 9.762 Å, $\alpha = 95.15°$, $\beta = 95.21°$ and $\gamma = 91.00°$, Z = 4, $D_{meas} = 3.27$ g/cm³. The structure, which comprises chains of thorium atoms linked by pairs of bridging sulfate groups, is illustrated in **Fig. 30** (p. 82). There are nine oxygen atoms around each thorium, four from bridging sulfate groups, two from a bidentate cyclic sulfate, one from a unidentate sulfate, and two from water molecules. The resulting co-ordination polyhedron is a trigonal prism with additional vertices over the square side faces. The Th-O bond distances (in Å) are in the range 2.30 to 2.58, viz: Th-O (H_2O), 2.54 and 2.58; Th-O (bidentate SO_4^{2-}), 2.49 and 2.46; Th-O (bidentate, bridging SO_4^{2-}), 2.49, 2.30 and 2.41, 2.35; Th-O (monodentate SO_4^{2-}) 2.34. The oxygen-oxygen distances fall in the range 2.71 to 3.33 Å, the two shortest distances, 2.71 and 2.72 Å, indicating hydrogen bonding between the water molecules and the oxygens of the sulfate groups.

The only other complexes for which X-ray data have been published are $(CN_3H_6)Na[Th(SO_4)_3] \cdot 6\,H_2O$, which crystallises in the hexagonal space group $P\bar{6}m2\text{-}D_{3h}^1$ (No. 187) with a = 9.5, c = 5.6 Å, Z = 1 [5] and $[Co(NH_3)_6]_2Th(SO_4)_5 \cdot 2\,H_2O$ for which X-ray powder diffraction results have been published but not interpreted [11].

In the absence of X-ray structural information on other thorium sulfato complexes Molodkin and co-workers have used the structure of $K_4Th(SO_4)_4 \cdot 2\,H_2O$ together with infrared, DTA, and TGA results to predict the numbers of water molecules present in the inner co-ordination spheres of the various hydrated sulfato complexes [4, 6, 8, 9, 15, 17]. Although the proposed structures may ultimately be confirmed the current evidence is not considered to be sufficiently reliable for them to be used in the present article.

     References for 10.6.8 on pp. 86/7

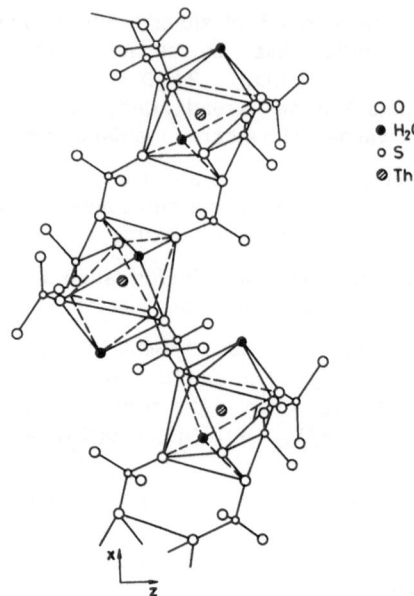

Fig. 30. Projection of the structure of $K_4[Th(SO_4)_4] \cdot 2 H_2O$ on the (010) plane [13].

Pycnometrically determined densities for certain of the sulfato complexes are listed in Table 46 (p. 80).

As expected for thorium(IV) compounds the sulfato complexes are all diamagnetic; χ_g and χ_{mol} values for $Na_2Th(SO_4)_3 \cdot 6 H_2O$, $Na_5Th(SO_4)_{4.5} \cdot 2.5 H_2O$, $K_4Th(SO_4)_4$, $K_2Th(SO_4)_3 \cdot 4 H_2O$, $K_2Th(SO_4)_3 \cdot 5 H_2O$, $K_4[Th(SO_4)_4(H_2O)_2]$, $Rb_2Th(SO_4)_3$, $Rb_2Th(SO_4)_3 \cdot 1.4 H_2O$, $Rb_4Th(SO_4)_4$, $Cs_2Th(SO_4)_3 \cdot 2 H_2O$, $Cs_4Th(SO_4)_4 \cdot H_2O$, $Cs_6Th(SO_4)_5 \cdot 3 H_2O$, $(NH_4)_2Th(SO_4)_3 \cdot 5 H_2O$, $(NH_4)_4Th(SO_4)_4 \cdot 2 H_2O$, $(NH_4)_6Th(SO_4)_5 \cdot 3 H_2O$, $(NH_4)_8Th(SO_4)_6 \cdot 2 H_2O$, and $(CN_3H_6)Na[Th(SO_4)_3(H_2O)_3] \cdot 3 H_2O$ are listed in [14]. Most values (including those for other Th compounds) showed poor reproducibility. Better reproducibility was obtained for $Na_2Th(SO_4)_3 \cdot 6 H_2O$ with $\chi_g = -0.314 \times 10^{-6}$ cm^3/g and $\chi_{mol} = -205 \times 10^{-6}$ cm^3/mol [14].

Infrared spectra have been recorded for several of the white sulfato complexes [15]. Tentative assignments are given in Tables 47, 48, and 49 (pp. 83, 84, and 85) for certain of the tris-, tetrakis-, pentakis-, and hexakis(sulfato) complexes. The spectra are in general very complex and in the absence of full-coordinate analyses it is not possible to be certain where the boundary between ν_1 and ν_3 modes is, for example, in complexes such as $Cs_2Th(SO_4)_3 \cdot 2 H_2O$, $Cs_4Th(SO_4)_4 \cdot H_2O$, and $(NH_4)_8Th(SO_4)_6 \cdot 2 H_2O$. It is likely that the complex spectra are a consequence in most instances of the presence of differently co-ordinated sulfate groups, as reported for $K_4Th(SO_4)_4 \cdot 2 H_2O$ (p. 81).

Raman spectra are reported in [16] for aqueous solutions of $(NH_4)_4Th(SO_4)_4$.

Available refractive indices for the thorium sulfato complexes are listed in Table 46 (p. 80).

Chemical Properties

With the exception of $K_2Th(SO_4)_3 \cdot 4 H_2O$, for which no specific statement is made, the complexes listed in Table 46 (p. 80) are reported to be air-stable at room temperature [4, 5,

8, 9, 11]. In contrast to the anhydrous rubidium tris- and tetrakis(sulfato) complexes, $K_4Th(SO_4)_4$ is converted to the monohydrate on exposure to the atmosphere [8]. Conversion of the various hydrates to anhydrous salts by heating them in air and by desiccation over concentrated sulfuric acid is dealt with above (p. 79). Although $K_4Th(SO_4)_4 \cdot 2\,H_2O$ readily loses water when washed with ether or methanol, which give, respectively, $K_4Th(SO_4)_4 \cdot H_2O$ and $K_4Th(SO_4)_4$ [8], like $(CN_3H_6)NaTh(SO_4)_3 \cdot 6\,H_2O$ it is stable over concentrated sulfuric acid [5].

Table 47
Infrared Data for Hydrated Tris(sulfato)thorates(IV) [15].

$K_2Th(SO_4)_3$ $\cdot 2\,H_2O$	$Cs_2Th(SO_4)_3$ $\cdot 2\,H_2O$	$(NH_4)_2Th(SO_4)_3$ $\cdot 5\,H_2O$	$(CN_3H_6)NaTh(SO_4)_3$ $\cdot 6\,H_2O$	assignment
	400 w, sh	400 w	400 w	$\nu_2(E)SO_4^{2-}$ [1]
	465 sh	452 m	428 w	
		470 m	460 m	
	495 w	525 w	530 m	$\nu_4(F_2)SO_4^{2-}$ [1],[2]
	517 w	590 s	597 m	
	598 m	624 sh	632 m	
	615 s	640 m		
	640 s			
650 s	658 m			
791 w	782 m	810 m		[3]
	854 w			
957 s	965 m	1005 s	1011 s	$\nu_1(A_1)SO_4^{2-}$
998 sh, w	1001 sh	1015 s	1046 s	
1020 sh	1016 sh	1059 s		
1041 s	1064 s			
	1080 s			[4]
1098 s	1098 s	1102 s	1116 s	$\nu_3(F_2)SO_4^{2-}$
1127 s	1115 s	1175 s	1210 s	
1144 sh	1145 s	1190 sh		
1220 s	1207 s	1203 sh		
		1435		νNH_4^+
		1470 s		
1626 m	1687 m	1640 m	1648 s	δH_2O
1654 sh		1670 sh, w	1674 s	
	3230 s, b	3080 s	3355 s	νOH, νNH [1]
		3230 s	3450 s	
		3460 s	3610 s	
		3530 sh, w		

[1] Region not studied for $K_2Th(SO_4)_3 \cdot 2\,H_2O$. — [2] It is not clear whether the bands below ca. 590 cm^{-1} should be attributed to ν_2 or ν_4 modes: compare assignments by same authors for tetrakis(sulfato) complexes in Table 48 (p. 84). — [3] Bands not assigned. — [4] As indicated in the text (p. 82) the boundary between ν_1 and ν_3 vibrations, particularly in the case of $Cs_2Th(SO_4)_3 \cdot 2\,H_2O$, is not clear in the absence of full-coordinate analyses.

Table 48
Infrared Data for Tetrakis(sulfato)thorates(IV) [15].

$K_4Th(SO_4)_4$ · 2 H_2O	$(NH_4)_4Th(SO_4)_4$ · 2 H_2O	$Cs_4Th(SO_4)_4$ · H_2O	$K_4Th(SO_4)_4$	$Rb_4Th(SO_4)_4$	assignment
440 sh	400 sh	420 sh	420 sh		
455 w	420 sh	460 sh	435 w		$\nu_2(E)SO_4^{2-}$
503 w	465 w	498 m	498 s		
	507 m		521 sh		
605 m	605 s	598 s	598 s		
615 m	617 s	616 s	616 s		$\nu_4(F_2)SO_4^{2-}$
630 w, sh	645 s	642 s	645 s		
640 s		660 w, sh	655 s	650 s	
810 w	810 w, b	774 w		832 m	
		863 w			1)
		927 m(?)			
963 s	964 s	960 s	960 s	960 s	
980 s	983 s	1001 s	1004 s	988 s	
1000 s	1010 s	1014 s	1020 s	1016 s	$\nu_1(A_1)SO_4^{2-}$
1021 s	1028 s	1035 sh	1044 s	1055 s	
1042 s	1040 w, sh	1055 s			
		1080 sh		1079 s	2)
		1090 s			
1100 s	1103 s	1105 s	1100 s	1130 s	
1120 s	1143 s	1135 s	1155 s	1212 s	
1150 s	1170 sh	1210 s	1227 s		
1176 s	1185 sh				$\nu_3(F_2)SO_4^{2-}$
1194 s	1223 s				
1212 s					
1240 s					
	1430 sh				νNH_4^+
	1460 s				
1660 w	1647 m	1645 w, sh	1620 w	1597 w	δH_2O 3)
		1685 m		1610 w	
3150 m	3100 w, sh	3250 m, b	3420 sh		
3400 sh	3250 s	3430 sh, w	3550 m		νNH, νOH 3),4)
	3580 sh		3625 m		

1) The 832 cm^{-1} band in the spectrum of $Rb_4Th(SO_4)_4$ (and one not listed at 1381) is attributed to NO_3^- impurity. The bands in this region of the spectra of the other compounds are not assigned. — 2) As indicated in the text the boundary between ν_1 and ν_3 vibrations in such complex spectra, particularly that of $Cs_4Th(SO_4)_4$ · H_2O, is not clear in the absence of full-coordinate analyses. — 3) These bands in the spectra of the anhydrous complexes are attributed to traces of adsorbed water. — 4) This region was not studied for $Rb_4Th(SO_4)_4$.

Table 49
Infrared Data for Pentakis- and Hexakis(sulfato)thorates(IV) [15].

$Cs_6Th(SO_4)_5$ $\cdot 3 H_2O$	$(NH_4)_6Th(SO_4)_5$ $\cdot 3 H_2O$	$(NH_4)_8Th(SO_4)_6$ $\cdot 2 H_2O$	assignment
390 w	400 sh	460 w	
425 w, sh	420 w, sh	498 w	$\nu_2(E)SO_4^{2-}$
495 m	460 w, sh		
	498 m		
516 w	532 w	601 s	
530 vw	605 sh	617 s	
603 s	618 s	647 s	$\nu_4(F_2)SO_4^{2-}$ [1]
617 s	646 s		
643 s			
775 sh			[2]
960 s	966 s	965	
968 sh	1015 w, sh	975 sh	$\nu_1(A_1)SO_4^{2-}$
1020 sh	1032 s	1017 s	
1037 s		1035 s	
		1070 s	[3]
1137 s	1127 s, b	1130 s, b	
1188 s	1180 sh	1201 s	$\nu_3(F_2)SO_4^{2-}$
1215 s	1211 s		
	1460 s, b	1430 sh	νNH_4^+
		1460 s, b	
1630 to 1680 w, b	1650 sh	1655 m	δH_2O
	1670 m		
3340 s, b	3080 s	3230 s	
3550 w, sh	3240 s	3540 m	$\nu OH, \nu NH$
	3460 s		
	3605 sh		

[1] It is not clear whether the bands below ca. 600 cm^{-1} should be attributed to ν_2 or ν_4 modes; compare assignments by same authors for tetrakis(sulfato) complexes in Table 48. — [2] Band not assigned. — [3] As indicated in the text (p. 82), the boundary between the ν_1 and ν_3 vibrations, particularly for $(NH_4)_8Th(SO_4)_6 \cdot 2H_2O$, is not clear in the absence of full-coordinate analyses.

The anhydrous ammonium complexes $(NH_4)_{2m-4}Th(SO_4)_m$, which form when the hydrates are heated at moderate temperatures in air (see Table 46, p. 80), are reported [4, 6] to decompose to $Th(SO_4)_2$ at ca. 550°C via the intermediate formation of $NH_4HTh(SO_4)_3$ at ca. 450°C. However, the TGA results show no plateaux and this phase has never been isolated. The ultimate decomposition product, at >750°C, for each of these complexes is ThO_2 [6]. The complex $(CN_3H_6)NaTh(SO_4)_3$, observed to form from the hexahydrate at ca. 170°C, is stated [5] to decompose to a mixture of Na_2SO_4 and ThO_2 via the intermediate $Na_2Th_2(SO_4)_5$ at around 500°C; again confirmation of the existence of this proposed phase is lacking. The anhydrous alkali metal sulfates are stable to temperatures in the range 650 to 700°C [4, 8, 9], whilst

References for 10.6.8 on pp. 86/7

$[Co(NH_3)_6]_2Th(SO_4)_5$ decomposes to the constituent sulfates at 450 to 750°C and finally to a mixture of Co_3O_8 and ThO_2 at ca. 900°C [11].

The sulfato complexes listed in Table 46 (p. 80), and $K_4Th(SO_4)_4$, are all water soluble, $Rb_2Th(SO_4)_3$ and $Cs_2Th(SO_4)_3 \cdot 2 H_2O$ are only sparingly so, and several are reported to dissolve in aqueous solutions of alkali metal and ammonium carbonates [4, 5, 9]. Information on the behaviour of $Na_2Th(SO_4)_3 \cdot 6 H_2O$ and $Na_2Th(SO_4)_3 \cdot 4 H_2O$ in the $Th(SO_4)_2$-Na_2SO_4-H_2O system at 25°C and 50°C is given in Tables 42 and 43 (pp. 71 and 72) and Figs. 25 and 26 (pp. 70 and 71).

The aqueous solutions of the sulfato complexes have a weakly acidic reaction (see Table 50) and also exhibit molar conductances higher than those expected for simple dissociation into cations and a complex anion (Table 51) [4, 5, 8, 9].

The complexes are insoluble in organic solvents such as methanol, ethanol, acetone, ether, and toluene [4, 5, 8, 9].

Table 50
pH of Aqueous Solutions of Sulfato Complexes of Thorium and Contribution of the Hydrogen Ions, μ_H, to the Molar Conductance [9].

compound	pH at dilution v in L/mol			μ_H in $\Omega^{-1} \cdot cm^2$
	v = 250	v = 500	v = 1000	at v = 250
$Na_2Th(SO_4)_3 \cdot 6 H_2O$	3.15	3.30	3.50	62
$K_2Th(SO_4)_3 \cdot 4 H_2O$	3.00	3.10	3.30	88
$Cs_2Th(SO_4)_3 \cdot 2 H_2O$	2.90	3.10	3.20	110
$K_4Th(SO_4)_4 \cdot 2 H_2O$	3.30	3.40	3.50	45
$Cs_4Th(SO_4)_4 \cdot H_2O$	3.25	3.40	3.50	49
$Cs_6Th(SO_4)_5 \cdot 3 H_2O$	3.35	3.50	3.60	39

Table 51
Conductivity Data for Hydrated Sulfato Complexes [9].

compound	Λ_{1000} in $\Omega^{-1} \cdot cm^2$ at 25°C	number of ions in solution	
		expected	observed
$(NH_4)_2Th(SO_4)_3 \cdot 5 H_2O$	470	3	4 to 5
$(NH_4)_4Th(SO_4)_4 \cdot 2 H_2O$	679	5	6
$(NH_4)_6Th(SO_4)_5 \cdot 3 H_2O$	930	7	8
$(NH_4)_8Th(SO_4)_6 \cdot 2 H_2O$	1200	9	10
$K_2Th(SO_4)_3 \cdot 4 H_2O$	466	3	4 to 5
$K_4Th(SO_4)_4 \cdot 2 H_2O$	701	5	6 to 7
$Cs_2Th(SO_4)_3 \cdot 2 H_2O$	475	3	4 to 5
$Cs_4Th(SO_4)_4 \cdot H_2O$	680	5	6
$Cs_6Th(SO_4)_5 \cdot 3 H_2O$	920	7	8

References for 10.6.8:

[1] J. W. Mellor (A Comprehensive Treatise on Inorganic and Theoretical Chemistry, Vol. VII, Longmans, London 1947, pp. 240/8). − [2] J. Flahaut (in: P. Pascal, Nouveau Traité

de Chimie Minérale, Vol. IX, Paris 1967, pp. 1103/9). − [3] Huang-Pang Chang, P. I. Fedorov (Zh. Neorgan. Khim. **6** [1961] 971/6; Russ. J. Inorg. Chem. **6** [1961] 495/7). − [4] A. K. Molodkin, G. A. Skotnikova, O. M. Ivanova (Zh. Neorgan. Khim. **10** [1965] 2243/53; Russ. J. Inorg. Chem. **10** [1965] 1220/6). − [5] A. K. Molodkin, G. A. Skotnikova, E. G. Arutyunyan (Zh. Neorgan. Khim. **9** [1964] 2705/9; Russ. J. Inorg. Chem. **9** [1964] 1458/61).

[6] A. K. Molodkin, G. A. Skotnikova, O. M. Ivanova (Zh. Neorgan. Khim. **11** [1966] 2241/4; Russ. J. Inorg. Chem. **11** [1966] 1201/9). − [7] V. M. Vdovenko, M. P. Koval'skaya, E. V. Shirvanskii (Radiokhimiya **3** [1961] 1/6; Radiochemistry [USSR] **3** [1961/64] 1/5). − [8] A. K. Molodkin, G. A. Skotnikova, O. M. Ivanova (Zh. Neorgan. Khim. **10** [1965] 2441/8; Russ. J. Inorg. Chem. **10** [1965] 1329/33). − [9] A. K. Molodkin, G. A. Skotnikova, O. M. Ivanova (Zh. Neorgan. Khim. **10** [1965] 2675/83; Russ. J. Inorg. Chem. **10** [1965] 1453/7). − [10] M. Hoshi, K. Ueno (J. Nucl. Sci. Technol. [Tokyo] **15** [1978] 585/8).

[11] K. Ueno, M. Hoshi (J. Inorg. Nucl. Chem. **33** [1971] 1765/73). − [12] E. G. Arutyunyan, M. A. Porai-Koshits (Zh. Struk. Khim. **4** [1963] 110/11; J. Struct. Chem. [USSR] **4** [1963] 96/7). − [13] E. G. Arutyunyan, M. A. Porai-Koshits, A. K. Molodkin (Zh. Struk. Khim. **7** [1966] 733/7; J. Struct. Chem. [USSR] **7** [1966] 683/6). − [14] V. I. Belova, Ya. K. Syrkin, A. K. Molodkin, O. M. Ivanova, L. M. Shiporina (Zh. Neorgan. Khim. **13** [1968] 1458/60; Russ. J. Inorg. Chem. **13** [1968] 766/7). − [15] O. N. Evstaf'eva, A. K. Molodkin, G. G. Dvoryantseva, O. M. Ivanova, M. I. Struchkova (Zh. Neorgan. Khim. **11** [1966] 1306/15; Russ. J. Inorg. Chem. **11** [1966] 697/702).

[16] L. V. Volod'ko, Lieh T'han Huoah (Zh. Prikl. Spektrosk. **10** [1969] 779/83; J. Appl. Spectrosc. [USSR] **10** [1969] 518/22). − [17] A. K. Molodkin, E. G. Arutyunyan (Zh. Neorgan. Khim. **10** [1965] 352/62; Russ. J. Inorg. Chem. **10** [1965] 189/95).

10.6.9 Thorium(IV) Mixed Acid Sulfato Complexes

Golovnya et al. [1] have reported the formation of $Na_4Th(SO_3)_{3.5}(SO_4)_{0.5} \cdot 7.5\,H_2O$, $Na_6Th(SO_3)_{3.5}(SO_4)_{1.5} \cdot 9\,H_2O$, and $(NH_4)_4H_4Th(SO_3)_5(SO_4) \cdot 12\,H_2O$ as a consequence of the partial oxidation of sulfite to sulfate during the preparation of thorium(IV) sulfato complexes. No preparative details or properties of products were given and further studies are required to establish whether these were genuine phases or mixtures.

Reference for 10.6.9:

[1] V. A. Golovnya, A. K. Molodkin, V. N. Tverdokhlebov (Zh. Neorgan. Khim. **10** [1965] 2196/8; Russ. J. Inorg. Chem. **10** [1965] 1195/6).

10.7 Thorium(IV) Fluorosulfate, $Th(SO_3F)_4$

$Th(SO_3F)_4$ is obtained by refluxing Th tetraacetate with an excess of fluorosulfuric acid, the white product being washed first with HSO_3F and then with SO_2Cl_2 before being dried in vacuo. Structural data are not available but infrared results have been tentatively interpreted as indicating eightfold co-ordination around the Th atom, with each SO_3F^- ion co-ordinated via two oxygen atoms. The observed IR modes are 1360 and 1170 (v_4, E), 1075 (v_1, A_1), 840 (v_2, A_1), 610 (v_5, E), 575, 545 (v_3, A_1), 450, 380 (v_6, E) cm^{-1}. $Th(SO_3F)_4$ is stable up to 160°C, above which temperature decomposition to $Th(SO_4)_2$ occurs, with evolution of SO_2F_2. It is a hygroscopic solid which reacts with donor ligands in CCl_4 or C_6H_6, in which it is insoluble, to give products of the types $Th(SO_3F)_4 \cdot 2\,L$ (L = pyridine, quinoline, dimethylsulfoxide, and triphenylphosphane oxide) and $Th(SO_3F)_4 \cdot L$ (L = bipyridine) [1].

Reference for 10.7:

[1] R. C. Paul, S. Singh, R. D. Verma (J. Indian Chem. Soc. **58** [1981] 24/5).

11   Compounds of Thorium and Selenium

Horst Wedemeyer
Kernforschungszentrum Karlsruhe
Institut für Material- und Festkörperforschung
Karlsruhe, Federal Republic of Germany

11.1   Binary Thorium Selenides

11.1.1   The Th-Se System

The binary compounds ThSe, Th_2Se_3, Th_7Se_{12}, $ThSe_2$, Th_2Se_5, and $ThSe_3$ are reported to exist in the thorium-selenium system, based on X-ray diffraction measurements and chemical analysis. A further compound, $ThSe_{2.33}$ (or Th_3Se_7), prepared by the reaction of the elements, could not be confirmed crystallographically [1].

There is a small range of solid solution of selenium in thorium [1], see also [2]. No intermediate phases were found to exist between thorium and ThSe, but a eutectic composition appears within the composition range of $ThSe_{0.1}$ and $ThSe_{0.2}$ at $1600 \pm 20\,°C$, extending over the range $ThSe_{0.02}$ to $ThSe_{0.8}$, approximately [1], see also [2]. The cubic ThSe (NaCl type), golden in color, has a small homogeneity range extending from $ThSe_{0.95}$ to $ThSe_{1.1}$ [1], see also [2]. The melting point of stoichiometric $ThSe_{1.0}$ was measured to be $1880\,°C$ [1], see also [2 to 4]. The melting point decreases for $ThSe_{0.9}$ to $1800\,°C$ and to $1680\,°C$ for $ThSe_{1.1}$ [1]. The orthorhombic Th_2Se_3 (Sb_2S_3 type), purple-black ($ThSe_{1.4}$) or gray-black ($ThSe_{1.5}$) in color, has a small homogeneity range extending from $ThSe_{1.4}$ to $ThSe_{1.5}$ [1], see also [2]. Th_2Se_3 decomposes peritectically at $1480 \pm 10\,°C$ [1], see also [2]. A eutectic is formed at $1460\,°C$, placed between Th_2Se_3 and Th_7Se_{12} [1], see also [2]. The hexagonal Th_7Se_{12} (Th_7S_{12} type), gray in color, melts at about $1460\,°C$, slightly above the eutectic Th_2Se_3-Th_7Se_{12} [1], see also [2, 3]. There remains some doubt whether the prepared Th_7Se_{12} is an oxygen-stabilized phase or not [5]. The orthorhombic $ThSe_2$ ($PbCl_2$ type), dark purplish gray in color, decomposes on heating to $1000\,°C$ [1], see also [2, 3]. Th_2Se_5 is tetragonal [6], see also [2], or orthorhombic (pseudo-tetragonal) [7], see also [2]. It is assumed that some range of homogeneity exists due to the differences in X-ray determination [7], see also [2]. The monoclinic $ThSe_3$ ($ZrSe_3$ type) is isostructural with $ThTe_3$, US_3, USe_3, and UTe_3 [7], see also [2].

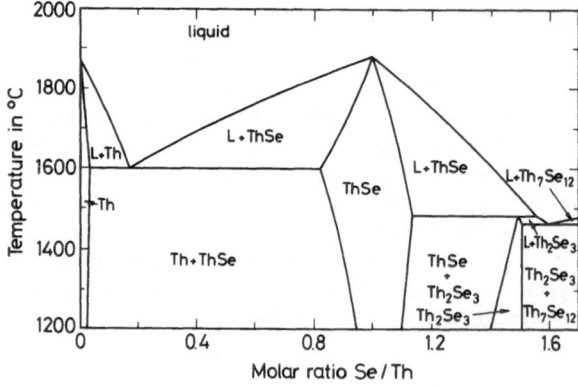

Fig. 31. Tentative phase diagram for the system thorium-selenium, from [1].

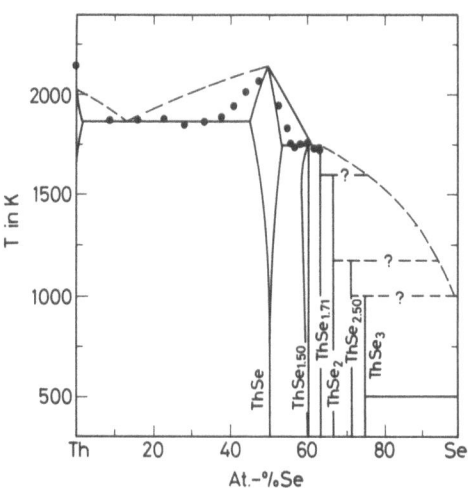

Fig. 32. Tentative phase diagram for the system thorium-selenium, from [2].

A tentative phase diagram, based on the measured melting points in the range of Th to Th$_7$Se$_{12}$ [1], is given in **Fig. 31**, see also [8, 9]. A further tentative phase diagram, including the compounds ThSe$_2$, Th$_2$Se$_5$, and ThSe$_3$ [2], is given in **Fig. 32**.

11.1.2   Thorium Monoselenide, ThSe

11.1.2.1   Formation and Preparation

Synthesis from the Elements

Thorium monoselenide, ThSe, was prepared for the first time in 1952 [1]. Stoichiometric amounts of thorium metal and selenium were sealed in an evacuated quartz tube and heated up slowly to 400°C to allow a fair quantity of selenium to be absorbed by the thorium metal. The strongly exothermic reaction started at 400°C. The temperature was then raised cautiously to 700°C. The samples could be handled in open air after cooling. Samples of monophasic ThSe with Se/Th ratios of 0.95 to 1.1 were prepared by this method [1], also cited in [2, 5, 10, 11]. The same reaction was carried out by preheating thorium filings (or powder [12]) and selenium in proper amounts sealed in an evacuated quartz tube at a temperature of 800°C. The reaction product was then ground (under 1 atm of argon [11]), pressed into pellets, sealed in a tantalum tube under 1 atm of argon, and sintered [11] or heated in an induction furnace [12] at 1850°C for 1 to 2 h. The interior of the sintered pellets was metallic golden in color, while the surface was shiny blue. Three phases were observed by X-ray determination: predominantly ThSe (NaCl phase), some ThOSe, and very little ThO$_2$. The monoselenides prepared had the nominal composition of Th$_{0.98}$Se and ThSe [11], see also [12]. This reaction was reported to be unsuccessful after heating the reaction partners for 2 weeks at 800°C, then grinding under inert conditions, sealing and re-evacuating, and again heating for 7 days to 800°C. No more than a small portion of the sample had reacted to form the monoselenide [6], also cited in [10].

References for 11.1 on p. 101

Reduction of $ThSe_2$

ThSe was obtained from the reaction of $ThSe_2$ with ThH_4; the product was ground and cold-pressed into pellets in a vacuum of 10^{-6} Torr at 1650°C [13].

Reaction of Thorium with ZnSe

ThSe was also prepared by the reaction of thorium metal powder and zinc selenide, ZnSe, in stoichiometric amounts. The reaction was carried out using a two stage procedure either in vacuum or in an inert atmosphere as the reaction was rather violent. The reaction took place at 800 to 900°C, and after vacuum sintering ThSe was obtained, dark in color with a greenish tinge. Oxygen impurities were measured to be about 0.46% [10].

Enthalpy and Gibbs Free Energy of Formation

The enthalpy of formation was estimated to be $\Delta H_f^\circ(ThSe, c, 298) = -85 \pm 10$ kcal/mol [14], see also [2]. The Gibbs free energy of formation is $\Delta G_f^\circ(ThSe, c, 298) = -84 \pm 10$ kcal/mol [2].

11.1.2.2   Crystallographic Properties, Bonding, and Lattice Dynamics

Thorium monoselenide is face-centered cubic (NaCl type) with $Z = 4$; the space group is O_h^5-Fm3m (No. 225) [1]. ThSe has a small homogeneity range extending from $ThSe_{0.95}$ to $ThSe_{1.1}$ (at 1200°C) with lattice constants varying from 5.879 kX (Th-rich) to 5.855 kX (Se-rich) (see **Fig. 33**) [1], see also [2]. Different measured values of the lattice constant, which are reported in the literature, are summarized in Table 52.

The X-ray density, calculated from diffraction measurements, is given as 10.20 g/cm³ [1], see also [17], 10.16 g/cm³ [19], see also [20], or 10.18 g/cm³ [4]. The calculated and observed densities at the phase limits are [1]: $D_{calc} = 9.90$ g/cm³, $D_{obs} = 9.90$ g/cm³ ($ThSe_{0.9}$), $D_{calc} = 9.54$ g/cm³, $D_{obs} = 9.26$ g/cm³ ($Th_{0.9}Se$). The densities were calculated on the assumption that the deviation from the stoichiometric composition ThSe is based on the presence of vacant selenium sites or thorium sites, respectively. The interatomic distances were calculated to be Th-Se $= 2.93$ kX and Th-Th $(=$ Se-Se$) = 4.14$ kX [1], or Th-Se $= 2.938$ Å [21]. The ionic radii, calculated (cation radius) and observed from cation-anion separation experiments are: anion radius $(Se^{2-}) = 1.953$ Å, cation radius $(Th^{4+}) = 0.985$ Å [21].

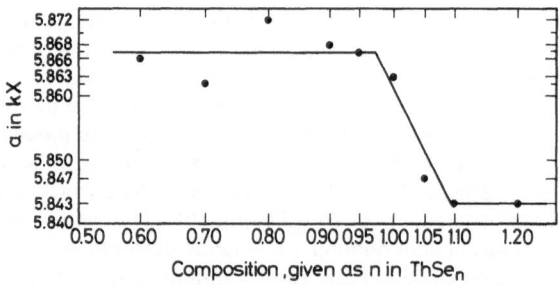

Fig. 33. Lattice constant a of the face-centered cubic phase of ThSe versus the gross composition of the sample [1].

Table 52
Measured Lattice Parameters of ThSe.

lattice constant a	method of preparation	Ref.; also referred in:
5.863 ± 0.002 kX	Th + Se	[1]; [15, 16]
5.879 kX (Th-rich)	Th + Se	[1]; [2, 17]
5.855 kX (Se-rich)	Th + Se	[1]; [2, 17]
5.875 Å	Th + Se	[6] from [1]; [4, 5, 10, 17]
5.785 Å	Th + ZnSe	[10]
5.879 ± 0.005 Å ($Th_{0.98}Se$)	Th + Se	[12]
5.875 ± 0.005 Å (ThSe)	Th + Se	[12]
5.869 ± 0.005 Å (ThSe)	Th + Se	[12]
0.5879 nm (ThSe)		[18]

The interatomic potentials for a large number of metal chalcogenide crystals, including ThSe, were calculated using a logarithmic interaction potential energy function. For ThSe, a cohesive energy of W = 3170 kJ/mol was obtained by a procedure which made use of the Moelwyn-Hughes parameter C_1 [22]. The following related values were calculated: compressibility $\beta_0 = 15.42 \times 10^{12}$ Pa^{-1}, force constant f = 11.4×10^{-4} N/m, IR absorption frequency $\nu_0 = 5.4 \times 10^{-12}$ Hz, Debye temperature $\Theta_D = 261$ K, Grüneisen parameter $\gamma = 1.80$, Anderson-Grüneisen parameter $\delta = 3.60$, Moelwyn-Hughes parameter $C_1 = 4.60$ [22]. Calculations within the framework of the Born model gave a cohesive energy of W = 832 kcal/mol when an exponential form was used, and W = 812 kcal/mol when an inverse power form was used. The zero point energy was 1.17 kcal/mol [23]. The following repulsive energies were calculated: 158 kcal/mol (exponential form) and 171 kcal/mol (inverse power form) for cation-anion interaction, 0.08 kcal/mol (exponential form) and 0.34 kcal/mol (inverse power form) for cation-cation interaction, and 10.9 kcal/mol (exponential form) and 17.5 kcal/mol (inverse power form) for anion-anion interaction [23].

The isothermal compressibility, K_T, of ThSe was measured with a high-pressure X-ray camera at pressures of up to 200 kbar resulting in $K_T = (7.55 \pm 0.7) \times 10^{-4}$ kbar^{-1} [24]. A NaCl-CsCl phase transition was not observed up to these pressures. This transition point is assumed to exist at 350 kbar [24].

Third-order elastic constants were calculated using the Born-Mayer potential model. The results for ThSe at 0 K were (in 10^{12} dyn/cm^2): $c^o_{111} = -22.248$, $c^o_{112} = -2.144$, $c^o_{123} = 0.844$ [25].

11.1.2.3 Thermal Properties

The melting point of the stoichiometric compound $ThSe_{1.0}$ was measured to be 1880°C [1], see also [2 to 4, 26, 27]. The melting point decreases for $ThSe_{0.9}$ to 1800°C at the thorium-rich phase boundary and for $ThSe_{1.1}$ to 1680°C at the selenium-rich phase boundary [1]. A further value for ThSe is reported to be 3160°F (1740°C) [19, 20].

The entropy of ThSe was estimated by a modified Latimer additivity scheme to be S°(ThSe, c, 298) = 22 ± 2 cal · mol^{-1} · K^{-1} [28], see also [2, 26]. Estimation comparing with the value for ThS gave 20 ± 2 cal · mol^{-1} · K^{-1} [14], see also [2]. Estimated values for the thermodynamic functions of gaseous ThSe are reported in [2]. The dissociation energy is

References for 11.1 on p. 101

Table 53
Thermodynamic Functions of Gaseous Thorium Monoselenide, ThSe(g) [2]. See note in Table 9 on p. 15.

T in K	C_p° in cal·mol⁻¹·K⁻¹	S° in cal·mol⁻¹·K⁻¹	$-(G_T^\circ - H_{298}^\circ)/T$ in cal·mol⁻¹·K⁻¹	$H_T^\circ - H_{298}^\circ$ in cal/mol	ΔH_f° in cal/mol	ΔG_f° in cal/mol	log K_p
298	8.629	63.180	63.180	0	84000	71979	−52.762
300	8.633	63.233	63.180	16	83993	71905	−52.382
400	8.769	65.736	63.520	886	83581	67936	−37.118
500	8.888	67.704	64.167	1768	81649	64093	−28.015
600	9.086	69.340	64.897	2666	81000	60643	−22.089
700	9.412	70.764	65.635	3590	80386	57300	−17.890
800	9.854	72.049	66.358	4553	79805	54042	−14.763
900	10.363	73.239	67.057	5564	79249	50856	−12.349
1000	10.873	74.357	67.732	6626	64946	47355	−10.349
1100	11.328	75.416	68.382	7737	64785	45605	− 9.061
1200	11.696	76.418	69.011	8889	64643	43867	− 7.989
1300	11.944	77.365	69.617	10072	64512	42141	− 7.085
1400	12.090	78.256	70.203	11274	64378	40426	− 6.311
1500	12.142	79.092	70.768	12486	64233	38719	− 5.641
1600	12.118	79.875	71.313	13700	64070	37024	− 5.057
1700	12.037	80.608	71.838	14908	63064	35373	− 4.547
1800	11.916	81.292	72.345	16106	62902	33748	− 4.098
1900	11.770	81.933	72.833	17290	62695	32135	− 3.696
2000	11.610	82.532	73.303	18459	62442	30533	− 3.336
2100	11.445	83.095	73.756	19612	58739	29073	− 3.026
2200	11.281	83.623	74.192	20748	58282	27670	− 2.749
2300	11.122	84.121	74.613	21868	57807	26289	− 2.498
2400	10.970	84.592	75.020	22973	57316	24928	− 2.270
2500	10.827	85.036	75.411	24063	56809	23590	− 2.062
2600	10.693	85.458	75.790	25139	56286	22272	− 1.872
2700	10.569	85.860	76.155	26202	55749	20973	− 1.698
2800	10.454	86.242	76.509	27253	55198	19695	− 1.537
2900	10.348	86.607	76.851	28293	54636	18438	− 1.390
3000	10.251	86.956	77.182	29323	54062	17201	− 1.253

assumed to be $D_0^o = 115 \pm 10$ kcal/mol and $D_{298}^o = 116 \pm 10$ kcal/mol [2]. From this, the enthalpy of formation is calculated to be $H_{f,298}^o = 89 \pm 10$ kcal/mol [2]. Further detailed data of the thermodynamic functions of gaseous ThSe are given in Table 53 [2].

11.1.2.4 Electrical Properties

The electric conductance of ThSe shows a metallic character [29], and the electric resistivity increases linearly with the temperature [12, 13], see also [11] (see **Fig. 34**). A value of the room temperature resistivity, measured at pressed powders (25000 psi), is 0.06 $\Omega \cdot$ cm [29].

Superconductivity was discovered in ThSe with three different samples, with the following transition temperatures: $T_C = 1.45$ K ($Th_{0.98}Se$), 1.72 K (ThSe), and 1.5 K (ThSe) [11, 12]. The transition temperature decreases linearly with the pressure at a rate of $dT/dp = -1.5 \times 10^{-5}$ K/bar [12].

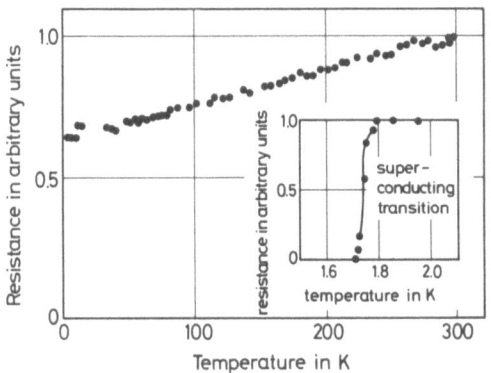

Fig. 34. Electrical resistance of ThSe versus temperature. The superconducting transition is displayed in the inset [12].

11.1.2.5 Magnetic Properties

ThSe shows a slightly paramagnetic behavior which is independent of temperature. The susceptibility was found to be 8×10^{-6} cm^3/mol [13]. From this, a spin fluctuation temperature of $T_{sf} = 50$ K was calculated [13].

11.1.2.6 Chemical Behavior

ThSe, golden in color, melts at 1880°C [1].

The thorium selenides are oxidized to ThO_2 and SeO_2 when heated up in air [30], and ignition occurs in a stream of oxygen [1]. They react with chlorine below 300°C forming $ThCl_4$ and $SeCl_4$ [30]. The alkali metals reduce the selenides to the metal, regardless of composition [30].

References for 11.1 on p. 101

A Th-ThSe eutectic is found to be placed between $ThSe_{0.1}$ and $ThSe_{0.2}$, melting at $1600 \pm 20\,°C$ [1].

From thorium- or thorium oxide-containing samples the selenides were leached out in concentrated HNO_3 [1]. The reaction of the chalcogenides with HNO_3 is vigorous [30]. All chalcogenides are attacked by boiling H_2SO_4; in 2 N H_2SO_4 some H_2Se is evolved [30].

11.1.3 Dithorium Triselenide, Th_2Se_3

11.1.3.1 Formation and Preparation

Synthesis from the Elements

Th_2Se_3 was synthesized for the first time in 1952 [1]. Stoichiometric amounts of thorium metal and selenium were sealed in an evacuated quartz tube and heated up slowly to $400\,°C$, to allow a fair quantity of selenium to be absorbed by the thorium metal. The strongly exothermic reaction started at $400\,°C$. The temperature was then raised cautiously to $700\,°C$. The samples could be handled in open air after cooling. Samples of Th_2Se_3 with Se/Th ratios of 1.4 to 1.5 were prepared by this method [1], see also [5]. This reaction was reported to be unsuccessful when heating 2 weeks at $800\,°C$, then grinding under inert conditions, sealing and re-evacuating, and again heating for 7 days to $800\,°C$ [6]. There remains some doubt whether the prepared Th_2Se_3 is an oxygen-stabilized phase or not [5].

Enthalpy and Gibbs Free Energy of Formation

The enthalpy of formation, measured by static bomb calorimetry, is reported to be $\Delta H_f^°(ThSe_{1.5}, c, 298) = -112$ kcal/mol [30], see also [2, 26]. The Gibbs free energy of formation was estimated to be $\Delta G_f^°(ThSe_{1.5}, c, 298) = -111 \pm 8$ kcal/mol [12].

11.1.3.2 Crystallographic Properties

Th_2Se_3 is orthorhombic (Sb_2S_3 type) with $Z = 4$; the space group is D_{2h}^{16}-Pnam (No. 62) [1], see also [16]. Th_2Se_3 was found to have a small homogeneity range extending from $ThSe_{1.4}$ (at least) to $ThSe_{1.5}$ (at $1200\,°C$) [1], see also [2]. The lattice parameters, measured by X-ray diffraction, are a = 11.32 ± 0.05, b = 11.55 ± 0.05, c = 4.26 ± 0.01 kX [1], see also [15], or a = 11.34, b = 11.57, c = 4.27 Å [6] from [1], see also [2, 5, 16, 17].

The atomic positions with the origin at 1 (space group Pnma) are Th(1): 4cm: $(x, 1/4, z)$, $(x, 3/4, z)$, $(1/2-x, 3/4, 1/2+z)$, $(1/2+x, 1/4, 1/2-z)$; Th(2): 4cm: $(x, 1/4, z)$, $(x, 3/4, z)$, $(1/2-x, 3/4, 1/2+z)$, $(1/2+x, 1/4, 1/2-z)$; Se(1): 4cm: $(x, 1/4, z)$, $(x, 3/4, z)$, $(1/2-x, 3/4, 1/2+z)$, $(1/2+x, 1/4, 1/2-z)$; Se(2): 4cm: $(x, 1/4, z)$, $(x, 3/4, z)$, $(1/2-x, 3/4, 1/2+z)$, $(1/2+x, 1/4, 1/2-z)$; Se(3): 4cm: $(x, 1/4, z)$, $(x, 3/4, z)$, $(1/2-x, 3/4, 1/2+z)$, $(1/2+x, 1/4, 1/2-z)$ [16].

11.1.3.3 Thermal Properties

Th_2Se_3 decomposes peritectically at $1480 \pm 10\,°C$ [1], also cited in [2, 3, 26].

A value of the entropy of Th_2Se_3 was estimated by a modified Latimer additivity scheme to be $S°(ThSe_{1.50}, c, 298) = 25.5 \pm 1$ cal $\cdot$ mol$^{-1} \cdot$ K^{-1} [28], see also [2, 26], and estimated from the estimated value of Th_2S_3 to be $S°(ThSe_{1.50}, c, 298) = 24.5 \pm 2.5$ cal $\cdot$ mol$^{-1} \cdot$ K^{-1} [14], see also [2].

11.1.3.4 Electrical Properties

Th$_2$Se$_3$ is reported to be a semiconductor (n-type) with a room temperature resistivity of 4.0 $\Omega \cdot$ cm (powders pressed at a pressure of 25000 psi) [29].

The thermoelectric power of Th$_2$Se$_3$ was measured on pressed powders (100000 psi) to be 0.1 mV/°C [29].

11.1.3.5 Chemical Behavior

Th$_2$Se$_3$, purple-black (ThSe$_{1.4}$) or gray-black (ThSe$_{1.5}$) in color, decomposes peritectically at 1480 $\pm$ 10°C [1], see also [2, 3].

The selenides are oxidized to ThO$_2$ and SeO$_2$ when heated up in air [30], and ignition occurs in a stream of oxygen [1]. The selenides react with chlorine below 300°C forming ThCl$_4$ and SeCl$_4$ [30]. The alkali metals reduce the selenide to the metal, regardless of composition [30].

The reaction of HNO$_3$ with the M$_2$Se$_3$ chalcogenides is vigorous and accompanied by flashes of light [30]. In concentrated HNO$_3$, the selenides were leached out from thorium- and thorium oxide-containing samples [1]. All chalcogenides are attacked by boiling H$_2$SO$_4$; in 2 N H$_2$SO$_4$ some H$_2$Se is evolved from Th$_2$Se$_3$ [30].

A eutectic melting at 1460°C (slightly below the melting point of Th$_7$Se$_{12}$) is found to be placed between Th$_2$Se$_3$ and Th$_7$Se$_{12}$ [1].

11.1.4 Heptathorium Dodecaselenide, Th$_7$Se$_{12}$

11.1.4.1 Formation and Preparation

Synthesis from the Elements

Th$_7$Se$_{12}$ (ThSe$_{1.7}$) was prepared for the first time in 1952 [1]. Stoichiometric amounts of thorium metal and selenium were sealed in an evacuated quartz tube and heated up slowly to 400°C to allow a fair quantity of selenium to be absorbed by the thorium metal. The strongly exothermic reaction started at 400°C. The temperature was then raised cautiously to 700°C. The samples could be handled in open air after cooling [1]. This reaction was reported to be unsuccessful after heating 2 weeks at 800°C, then grinding under inert conditions, sealing and re-evacuating, and again heating for 7 days to 800°C [6]. Only the formation of ThSe$_2$ and a second, previously unreported phase (but no Th$_7$Se$_{12}$) was confirmed by X-ray diffraction measurement [6]. There remains some doubt whether the prepared Th$_7$Se$_{12}$ is an oxygen-stabilized phase or not [5].

Enthalpy and Gibbs Free Energy of Formation

The enthalpy of formation was estimated to be ΔH_f°(ThSe$_{1.7}$, c, 298) $= -120 \pm 20$ kcal/mol [14], see also [2]. The Gibbs free energy of formation was estimated to be ΔG_f°(ThSe$_{1.7}$, c, 298) $= -119 \pm 20$ kcal/mol [2].

11.1.4.2 Crystallographic Properties

Th$_7$Se$_{12}$ is hexagonal (Th$_7$Se$_{12}$ type) with Z = 1; the space group is C_{6h}^2-P6$_3$/m (No. 176) [31], see also [16]. The lattice parameters, determined from X-ray diffraction measurements, are a = 11.546 ± 0.006 kX, c = 4.22 ± 0.01 kX [31], see also [1]; or a = 11.58 Å, c = 4.36 Å [6] from [1]; a = 11.57 Å, c = 4.23 Å [5] from [1, 31]; a = 11.596 Å, c = 4.23 Å [16], see also [17]. The theoretical density was calculated to be 8.7 g/cm³ [31], see also [17]. The atomic positions are 1 Th(1) in: $\pm (0,0,{}^1/_4)$; 6 Th(2), 6 Se(1), and 6 Se(2) in: $\pm (x, y, {}^1/_4)$, $(\bar{y}, x-y, {}^1/_4)$, $(y-x, \bar{x}, {}^1/_4)$ with parameters 6 Th(2): x = 0.15, y = −0.28; 6 Se(1): x = 0.51, y = 0.38; 6 Se(2): x = 0.24, y = 0.0 [32], see also [16, 17]. The determined space group of $P6_3/m\text{-}C_{6h}^2$ (No. 176) does not strictly apply to this structure. Since there is only one Th(1) atom but two equivalent a sites $((0,0,{}^1/_4), (0,0,{}^3/_4))$ the Th(1)-Se(2) distances are different depending on whether the $(0,0,{}^1/_4)$ or the $(0,0,{}^3/_4)$ position is occupied (compare to Th_7S_{12} [32], in Section 10.1.4.2, p. 29). The interatomic distances are given as Th(1)-3 Se(2) = 3.12 kX, Th(1)-6 Se(2) = 3.20 kX, mean value: Th(1)-Se(2) = 3.17 kX [31].

11.1.4.3 Thermal Properties

The melting point of Th_7Se_{12} is 1460°C [1], also cited in [3].

The entropy of Th_7Se_{12} was estimated to be $S°(ThSe_{1.71}, c, 298) = 27 \pm 3$ cal · mol⁻¹ · K⁻¹ [28], see also [2, 26].

11.1.4.4 Electrical Properties

Th_7Se_{12} is reported to be a semiconductor (n-type) with a room temperature resistivity of 400 Ω · cm (powders pressed at 25000 psi) [29].

The thermoelectric power of Th_7Se_{12} was measured on pressed powders (100000 psi) to be 0.25 mV/°C [29].

11.1.4.5 Chemical Behavior

Th_7Se_{12}, gray in color, melts at 1460°C [1], see also [2]. The selenides are oxidized to ThO_2 and SeO_2 when heated in air [30], and ignition occurs in a stream of oxygen [1]. The selenides react with chlorine below 300°C forming $ThCl_4$ and $SeCl_4$ [30]. The alkali metals reduce the selenides to the metals, regardless of composition [30]. In concentrated HNO_3, the selenides were leached out from thorium- and thorium oxide-containing samples [1]. They are attacked by boiling H_2SO_4 [30]. A eutectic is found to be placed between Th_2Se_3 and Th_7Se_{12}, melting at 1460°C, slightly below the melting point of Th_7Se_{12} [1].

11.1.5 Thorium Diselenide, $ThSe_2$

11.1.5.1 Formation and Preparation

The preparation of a thorium (di)selenide was reported for the first time in 1896, when thorium carbide was heated with selenium vapor at elevated temperatures [33]. In 1905 it was prepared from the reaction of $ThCl_4$ with selenium vapor and from the reaction of $ThBr_4$ with H_2Se at red heat [34]. These compounds were not well defined (also see "Thorium" 1955, p. 294). The earliest study of a well characterized $ThSe_2$ was presented in 1952 [1].

Synthesis from the Elements

Thorium diselenide, ThSe$_2$, was prepared by reacting stoichiometric amounts of thorium metal and selenium in an evacuated and sealed quartz tube. The reaction mixture was heated up slowly to 400 °C to allow a fair quantity of selenium to be absorbed by the thorium metal. The strongly exothermic reaction started at 400 °C. The temperature was then raised cautiously to 700 °C. The samples could be handled in open air after cooling [1], also cited in [12]. A polyselenide, Th$_2$Se$_5$, was prepared in a first step, when the reaction was carried out with selenium in excess and sealed in a silica tube at 600 °C for 2 to 3 weeks to ensure complete reaction and reasonable crystal growth. The reduction of this Th$_2$Se$_5$ in vacuum at 900 °C led after several hours to ThSe$_2$ [6], see also [30].

Enthalpy and Gibbs Free Energy of Formation

Estimated values are reported for the enthalpy of formation to be $\Delta H_f^\circ(\text{ThSe}_2, c, 298) = -130 \pm 20$ kcal/mol and for the Gibbs free energy of formation to be $\Delta G_f^\circ(\text{ThSe}_2, c, 298) = -129 \pm 20$ kcal/mol [14], see also [2].

11.1.5.2 Crystallographic Properties

ThSe$_2$ is orthorhombic (PbCl$_2$ type) with $Z = 4$; the space group is Pmnb-D$_{2h}^{16}$ (No. 62) [31], see also [1, 16]. The lattice parameters, measured by X-ray diffraction on powdered samples, are a = 4.98 ± 0.01, b = 7.50 ± 0.01, c = 9.38 ± 0.01 kX [1], see also [15], or a = 4.99, b = 7.52, c = 9.40 Å [6] from [1]; a = 4.411 ± 0.002, b = 7.595 ± 0.002, c = 9.046 ± 0.002 kX [31] or a = 4.420, b = 7.611, c = 9.064 Å [2, 5, 16, 17] from [31]; a = 4.435, b = 7.629, c = 9.085 Å [6], see also [2, 12]. From this, a theoretical density of 8.5 g/cm^3 was calculated, as compared to a measured value of 8.2 g/cm^3 [31], see also [17].

The atomic positions for one set of four thorium atoms and two sets of four selenium atoms are $\pm$ ($^1/_4$, y, z); ($^1/_4$, $^1/_2$+y, $^1/_2$−z) [1], see also [16], with parameters Th: y = 0.25, z = 0.125, Se(1): y = 0.38, z = 0.43 [31], see also [17]; Se(2): y = 0.47, z = 0.82. Each thorium is co-ordinated to nine selenium atoms with an average distance of Th-Se = 3.08 kX; the individual distances are Th-2 Se(1) = 2.98 kX, Th-1 Se(2) = 3.22 kX, Th-1 Se(1) = 2.93 kX, Th-2 Se(2) = 3.09 kX, Th-1 Se(1) = 2.85 kX, Th-2 Se(2) = 3.27 kX [31].

11.1.5.3 Thermal Properties

ThSe$_2$ loses selenium when heated in vacuum to 1000 °C [1], see also [3]; it decomposes before melting [3], see also [1].

The standard entropy, S°, was estimated by a Latimer additivity scheme to be $S°(\text{ThSe}_2, c, 298) = 30 \pm 3$ cal · mol^{-1} · K^{-1} [28], see also [2, 26].

For gaseous ThSe$_2$ the atomization energy was estimated to be $D_{at, 298}^\circ = 210 \pm 20$ kcal/mol and the enthalpy of formation to be $\Delta H_{f, 298} = 46 \pm 20$ kcal/mol [2]. Further values of the thermodynamic functions are summarized in Table 54, p. 98 [2].

References for 11.1 on p. 101
7

Table 54
Thermodynamic Functions of Gaseous Thorium Diselenide, $ThSe_2(g)$ [2]. See note in Table 9 on p. 15.

T in K	C_p° in cal·mol⁻¹·K⁻¹	S° in cal·mol⁻¹·K⁻¹	$-(G_T^\circ - H_{298}^\circ)/T$ in cal·mol⁻¹·K⁻¹	$H_T^\circ - H_{298}^\circ$ in cal/mol	ΔH_f° in cal/mol	ΔG_f° in cal/mol	log K_p
298	13.243	78.070	78.070	0	46000	32552	−23.861
300	13.250	78.152	78.070	25	45991	32469	−23.653
400	13.524	82.005	78.594	1364	45431	28043	−15.322
500	13.659	85.038	79.590	2724	41850	23801	−10.403
600	13.734	87.535	80.712	4094	40833	20287	−7.389
700	13.780	89.656	81.842	5470	39860	16942	−5.289
800	13.810	91.498	82.937	6849	38897	13731	−3.751
900	13.831	93.126	83.980	8231	37911	10646	−2.585
1000	13.846	94.584	84.969	9615	9351	6922	−1.513
1100	13.857	95.904	85.904	11000	8996	6698	−1.331
1200	13.865	97.110	86.788	12387	8619	6505	−1.185
1300	13.872	98.221	87.626	13773	8219	6344	−1.067
1400	13.877	99.249	88.420	15161	7796	6216	−0.970
1500	13.882	100.206	89.174	16549	7350	6118	−0.891
1600	13.885	101.102	89.892	17937	6882	6053	−0.827
1700	13.888	101.944	90.576	19326	5574	6047	−0.777
1800	13.890	102.738	91.230	20715	5119	6089	−0.739
1900	13.892	103.489	91.856	22104	4630	6157	−0.708
2000	13.894	104.202	92.455	23493	4108	6251	−0.683
2100	13.896	104.880	93.031	24883	151	6500	−0.676
2200	13.897	105.526	93.584	26272	−546	6817	−0.677
2300	13.898	106.144	94.117	27662	−1246	7167	−0.681
2400	13.899	106.735	94.630	29052	−1948	7547	−0.687
2500	13.900	107.303	95.126	30442	−2652	7958	−0.696
2600	13.901	107.848	95.605	31832	−3360	8398	−0.706
2700	13.901	108.373	96.068	33222	−4070	8862	−0.717
2800	13.902	108.878	96.517	34612	−4784	9354	−0.730
2900	13.903	109.366	96.951	36002	−5498	9874	−0.744
3000	13.903	109.837	97.373	37393	−6215	10418	−0.759

11.1.5.4 Electrical Properties

ThSe$_2$ is a semiconductor (p-type) with a room temperature resistivity of 150000 $\Omega \cdot$ cm (powders pressed at 25000 psi) [29].

The thermoelectric power of ThSe$_2$ was measured to be -0.5 mV/°C (powders pressed at 100000 psi) [29].

11.1.5.5 Chemical Behavior

ThSe$_2$, dark purplish gray in color, decomposes on heating to 1000°C [1], see also [2, 3].

The selenides are oxidized to ThO$_2$ and SeO$_2$ when heated up in air [30]. The selenides react with chlorine below 300°C forming ThCl$_4$ and SeCl$_4$ [30]. The alkali metals reduce the selenides to the metal, regardless of composition [30].

The reaction of HNO$_3$ with the dichalcogenides is vigorous [30]. In concentrated HNO$_3$ the selenides were leached out from thorium- and thorium oxide-containing samples [1].

11.1.6 Dithorium Pentaselenide, Th$_2$Se$_5$

There is some evidence for the existence of a Th$_3$Se$_7$ (ThSe$_{2.33}$) polyselenide which was prepared both by synthesis from the elements at temperatures up to 700°C in sealed and evacuated quartz tubes and by tensimetric degradation of a sample of gross composition ThSe$_3$, when the sample was heated at 200°C in vacuum of 10^{-4} Torr. The X-ray diffraction patterns, however, were very complex and the existence of a Th$_3$Se$_7$ compound could not be confirmed crystallographically [1]. A compound of composition Th$_3$Se$_7$ was not confirmed from degradation experiments, and instead of this a compound Th$_2$Se$_5$ was obtained with a tetragonal structure [6]. Large differences in the cell dimensions may indicate a range of homogeneity for the Th$_2$Se$_5$ phase [2, 7].

11.1.6.1 Formation and Preparation

Synthesis from the Elements

Th$_2$Se$_5$ was obtained by the reaction of thorium metal with excess selenium. The reaction was carried out in an evacuated and sealed quartz tube at 600°C within 2 to 3 weeks to ensure complete reaction and reasonable crystal growth. The excess selenium was then sublimed off by heating up to 350°C leaving Th$_2$Se$_5$; no further selenium was driven off after one week at 350°C [6].

Enthalpy and Gibbs Free Energy of Formation

Estimated values, obtained by analogy with the neighboring compounds, are ΔH_f°(Th$_2$Se$_5$,c,298) $= -132 \pm 20$ kcal/mol, and ΔG_f°(Th$_2$Se$_5$,c,298) $= -131 \pm 20$ kcal/mol [2].

11.1.6.2 Crystallographic Properties

Th$_2$Se$_5$ is tetragonal with Z = 2, the space group being P4$_2$/nmc-D$_{4h}^{15}$ (No. 137) [6], see also [2], or orthorhombic (pseudotetragonal) with Z = 4; the space group is then Pcnb-D$_{2h}^{14}$ (No. 60)

[7], see also [2]. The measured lattice parameters for the tetragonal lattice are a = 5.629 Å, c = 10.764 Å [6], see also [2, 17]; a = 5.623 Å, c = 10.712 Å are reported in [2]; and for the orthorhombic (pseudo-tetragonal) lattice a = 7.94 ± 0.01 Å, c = 10.728 ± 0.005 Å [7], see also [2]. One sample was found with significantly different lattice dimensions a = 8.02 Å, c = 10.631 Å [7], see also [2], from which some nonstoichiometry was assumed for Th_2Se_5 [7], see also [2]. The theoretical density was calculated for the tetragonal cell to be 8.34 g/cm³, as compared to the measured density of 8.21 g/cm³ [6], see also [17].

11.1.6.3 Thermal Properties

Th_2Se_5 loses selenium when heated in vacuum to 950°C [30], see also [6]. It decomposes before melting [30], see also [6].

The standard entropy, S°, was estimated by a Latimer additivity scheme to be $S°(ThSe_{2.5}, c, 298) = 35 \pm 3$ cal · mol⁻¹ · K⁻¹ [28], see also [2, 26].

11.1.6.4 Electrical Properties

Th_2Se_5 is a semiconductor (p-type) with a room temperature resistivity of 10000 Ω · cm (powders pressed at 25000 psi) [29].

The thermoelectric power of Th_2Se_5 was measured to be −0.8 mV/°C (powders pressed at 100000 psi) [29].

11.1.6.5 Chemical Behavior

Th_2Se_5, dark purple in color, decomposes on heating to 950°C in vacuum [30].

The selenides are oxidized to ThO_2 and SeO_2 when heated up in air [30]. They react with chlorine below 300°C forming $ThCl_4$ and $SeCl_4$ [30]. The alkali metals reduce all selenides to the metal, regardless of composition [30].

The reaction of HNO_3 with the polychalcogenides is vigorous [30]. Th_2Se_5 was observed to react with 50 vol% H_2O_2 if boiling, but not at room temperature [30].

11.1.7 Thorium Triselenide, $ThSe_3$

11.1.7.1 Formation and Preparation

The preparation of $ThSe_3$ was published for the first time in 1980 [7], see also [2]. The reaction was carried out with thorium metal and selenium in large excess, placed in an evacuated and sealed quartz tube, at 600°C for one week [7].

An estimated value of the enthalpy of formation is $\Delta H_f^°(ThSe_3, c, 298) = -135 \pm 20$ kcal/mol [2].

11.1.7.2 Crystallographic Properties

ThSe$_3$ crystals are monoclinic (ZrSe$_3$ type) with Z = 2; the space group is P2$_1$/m-C$_{2h}^2$ (No. 11) [7]. The lattice parameters, obtained from X-ray diffraction, are a = 5.72 $\pm$ 0.01 Å, b = 4.21 $\pm$ 0.01 Å, c = 9.64 $\pm$ 0.01 Å, β = 97.05° $\pm$ 0.05° [7], see also [2], or a = 5.723 Å, b = 4.218 Å, c = 9.673 Å, β = 97.2° as reported in [2]. The theoretical density was calculated to be 6.77 g/cm^3 as compared to a measured value of 6.69 g/cm^3 [7].

11.1.7.3 Thermal Properties

The standard entropy, S°, was estimated using a modified Latimer additivity scheme to be S°(ThSe$_3$, c, 298) = 40 $\pm$ 4 cal · mol^{-1} · K^{-1} [2], see also [28].

References for 11.1:

[1] R. W. M. D'Eye, P. G. Sellman, J. R. Murray (J. Chem. Soc. **1952** 2555/62). – [2] F. Grønvold, J. Drowart, E. F. Westrum Jr. (in: F. L. Oetting, The Chemical Thermodynamics of Actinide Elements and Compounds, IAEA, Vienna 1984, pp. 27/32). – [3] R. W. M. D'Eye, P. G. Sellman (J. Chem. Soc. **1954** 3760/6). – [4] J. H. Handwerk, O. L. Kruger (Nucl. Eng. Design **17** [1971] 397/408). – [5] R. M. Dell, N. J. Bridger (MTP [Med. Tech. Publ. Co.] Intern. Rev. Sci. Inorg. Chem. Ser. One **7** [1972] 211/74).

[6] J. Graham, F. K. McTaggart (Australian J. Chem. **13** [1960] 67/73). – [7] H. Noel (J. Inorg. Nucl. Chem. **42** [1980] 1715/7). – [8] A. A. Bauer, F. A. Rough (Progr. Nucl. Energy Ser. V **2** [1959] 612/20). – [9] T. A. Badaeva (Str. Splavov Nek. Sist. Uranom Toriem **1961** 339/57 [The Structure of Alloys of Certain Systems Containing Uranium and Thorium]; AEC-TR-5834 [1963] 321/36; N.S.A. **16** [1962] No. 30877). – [10] G. H. B. Lovell, D. R. Perels, E. J. Britz (J. Nucl. Mater. **39** [1971] 303/10).

[11] A. R. Moodenbaugh (Diss. Univ. California 1975, pp. 1/169; N.S.A. **33** [1976] No. 30308). – [12] A. R. Moodenbaugh, D. C. Johnston, R. Viswanathan, R. N. Shelton, L. E. De Long, W. A. Fertig (J. Low Temp. Phys. **33** [1978] 175/203). – [13] M. Haessler, C. H. De Novion (J. Phys. C **10** [1977] 589/602). – [14] K. C. Mills (Thermodynamic Data for Inorganic Sulphides, Selenides and Tellurides, Butterworth, London 1974). – [15] M. Allbutt, R. M. Dell (J. Nucl. Mater. **24** [1967] 1/20).

[16] D. J. Lam, J. B. Darby Jr., M. V. Nevitt (Actinides Electron. Struct. Relat. Prop. **2** [1974] 119/84). – [17] K. Girgis (At. Energy Rev. Spec. Issue No. 5 [1975] 191/238). – [18] T. Palewski (Phys. Status Solidi A **84** [1984] K47/K50). – [19] M. S. Farkas, A. A. Bauer, R. F. Dickerson (BMI-1568 [1962] 1/20; C.A. **56** [1962] 13732). – [20] N. M. Griesenauer, M. S. Farkas, F. A. Rough (BMI-1680 [1964] 1/32; C.A **62** [1965] 2454).

[21] M. Allbutt, R. M. Dell (J. Inorg. Nucl. Chem. **30** [1968] 705/10). – [22] K. P. Thakur (Australian J. Phys. **30** [1977] 325/34). – [23] P. S. Bakhshi, V. K. Jain, J. Shanker (J. Inorg. Nucl. Chem. **43** [1981] 901/9). – [24] K. G. Rajan, R. Krishnan, A. Sequeira, G. Venkataraman (Proc. Nucl. Phys. Solid State Phys. Symp. C **16** [1973] 160). – [25] K. P. Thakur (J. Phys. Chem. Solids **41** [1980] 465/72).

[26] M. H. Rand (At. Energy Rev. Spec. Issue No. 5 [1975] 7/86). – [27] O. von Goldbeck (At. Energy Rev. Spec. Issue No. 5 [1975] 87/142). – [28] E. F. Westrum Jr., F. G. Grønvold (Thermodyn. Nucl. Mater. Proc. Symp., Vienna, 1962, pp. 23/36; SM-26/30 [1963] 1/15). – [29] F. K. McTaggart (Australian J. Chem. **11** [1958] 471/80). – [30] J. Bear, F. K. McTaggart (Australian J. Phys. **11** [1958] 458/70).

[31] R. W. M. D'Eye (J. Chem. Soc. **1953** 1670/2). – [32] W. H. Zachariasen (Acta Cryst. **2** [1949] 288/91). – [33] H. Moissan, A. Étard (Compt Rend. **122** [1896] 573/7; Bull. Soc. Chim. France **15** [1896] 1271/5). – [34] H. Moissan, Martinsen (Compt. Rend. **140** [1905] 1510/5).

11.2 Compounds of Thorium with Selenium and Oxygen

In this chapter only the one existing thorium oxide selenide, ThOSe, is described. The information on selenites and selenates is given in Chapter 11.4, p. 104.

11.2.1 Thorium Oxide Selenide, ThOSe

11.2.1.1 Preparation

ThOSe was prepared by the reaction of stoichiometric amounts of $ThSe_2$ and ThO_2, placed in an aluminium crucible and sealed in an evacuated quartz tube at 1000°C for 3 days. Pure ThOSe was obtained with a total amount of impurities of 0.3 at.% per molecule of the compound [1]. Single crystals of ThOSe were obtained from the elements (Th, Se) mixed with a third element (Si, Ge, P, As, Sb, or Bi) by a transport method with bromine or iodine as transporting agent (3 to 6 mg/cm³). The reactants are placed in uncoated quartz ampules heated to 900 to 1050°C at one end and 50 to 100°C cooler at the other, for 1 week. With this method the SiO_2 is reduced to elemental silicon (which may react with the third element to form SiP, SiAs, ...) and the oxygen is used for the formation of ThOSe. The crystals were purplish black in color (brownish red when powdered), most of which had a pseudo-hexagonal habit [2], also cited in [3].

11.2.1.2 Crystallographic Properties

ThOSe is tetragonal (PbFCl type) with Z = 2; the space group is P4/nmm-D_{4h}^7 (No. 129) [4], see also [2, 3, 5]. The lattice parameters, measured by X-ray diffraction, are a = 4.030 ± 0.005 kX, c = 7.005 ± 0.005 kX [4], see also [5, 6], or a = 4.04 Å, c = 7.02 Å [3], from [4]. From this, the X-ray density was calculated to be 9.49 g/cm³ as compared to the measured density of 9.61 g/cm³ [4]. The atomic positions for the PbFCl structure are 2 Th in $(^1/_2, 0, x)$, $(0, ^1/_2, \bar{x})$; 2 O in $(0, 0, 0)$, $(^1/_2, ^1/_2, 0)$; 2 Se in $(^1/_2, 0, z)$, $(0, ^1/_2, \bar{z})$, with parameters x = 0.18 ± 0.005 and z = 0.63 ± 0.01 [4]. The interatomic distances are Th-4 O = 2.28 kX and Th-5 Se = 3.15 kX [4]. The coordination polyhedron is shown in **Fig. 35**.

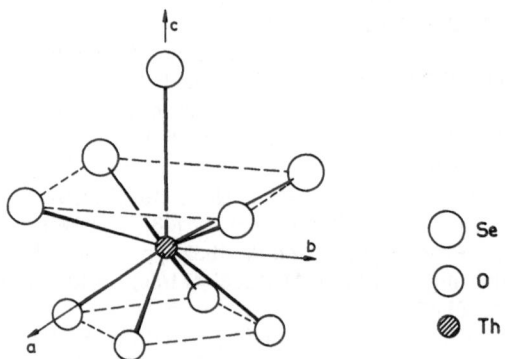

Fig. 35. Coordination polyhedron for ThOSe [1].

11.2.1.3 Thermal Properties

ThOSe melts without degradation at 2200°C [7].

The heat capacity, C_p, of ThOSe was measured within the temperature range of 5 to 300 K. The obtained values of C_p and calculated values for the entropy, S, are summarized in Table 55. The characteristic room temperature values are C_p = 72.65 J · mol^{-1} · K^{-1} (at 298.15 K), S = 93.50 J · mol^{-1} · K^{-1} (at 298.15 K) [1].

The total heat capacity is given as $C_p = C_{dil} + C_{latt} + C_{lin}$. Magnetic contributions are absent as ThOSe is not magnetic [1], or shows only a weak para- or diamagnetic behavior [2]. The dilation term, C_{dil}, according to the thermal lattice expansion, was assumed to be $C_{dil} = A \cdot C_p^2 \cdot T$ (with A = constant), the lattice contribution, C_{latt}, was calculated from Debye functions, and the electronic contribution, $C_{lin} = \gamma \cdot T$ is very small. The constants were calculated to be $A(300) = 0.25 \times 10^{-5}$, $\Theta_D(0) = 368$ K, $\Theta_D(300) = 400$ K, and $\gamma(0) = 1.0$ mJ · mol^{-1} · K^{-2} [1].

Table 55
Thermodynamic Functions for ThOSe [1].

T in K	C_p in J · mol^{-1} · K^{-1}	S in J · mol^{-1} · K^{-1}	T in K	C_p in J · mol^{-1} · K^{-1}	S in J · mol^{-1} · K^{-1}
5.0	0.02	0.01	140.0	53.14	45.40
10.0	0.17	0.05	150.0	55.28	49.14
15.0	0.81	0.22	160.0	57.15	52.77
20.0	2.13	0.62	170.0	58.87	56.29
25.0	4.08	1.29	180.0	60.58	59.70
30.0	6.53	2.24	190.0	62.17	63.02
35.0	9.36	3.46	200.0	63.53	66.24
40.0	12.41	4.90	210.0	64.88	69.37
45.0	15.57	6.55	220.0	65.96	72.42
50.0	18.72	8.35	230.0	67.03	75.37
60.0	24.71	12.30	240.0	67.94	78.25
70.0	29.93	16.52	250.0	68.84	81.04
80.0	34.46	20.81	260.0	69.64	83.76
90.0	38.47	25.11	270.0	70.45	86.40
100.0	42.07	29.35	280.0	71.33	88.98
110.0	45.32	33.52	290.0	72.13	91.49
120.0	48.21	37.59	300.0	72.75	93.95
130.0	50.79	41.55			

References for 11.2:

[1] G. Amoretti, A. Blaise, J. M. Collard, R. O. A. Hall, M. J. Mortimer, R. Troc (J. Magn. Magn. Mater. **46** [1984] 57/67). — [2] H. U. Boelsterli, F. Hulliger (J. Mater. Sci. **3** [1968] 664/5). — [3] R. M. Dell, N. J. Bridger (MTP [Med. Tech. Publ. Co.] Intern. Rev. Sci. Inorg. Chem. Ser. One **7** [1972] 211/74). — [4] R. W. M. D'Eye, P. G. Sellman, J. R. Murray (J. Chem. Soc. **1952** 2555/62). — [5] K. Girgis (At. Energy Rev. Spec. Issue No. 5 [1975] 191/238).
[6] M. Allbutt, R. M. Dell (J. Nucl. Mater. **24** [1967] 1/20). — [7] R. W. M. D'Eye, P. G. Sellman (J. Chem. Soc. **1954** 3760/3).

11.3   Compounds of Thorium with Selenium and Nitrogen

There is only one compound known within the system thorium-selenium-nitrogen: Th_2N_2Se.

11.3.1   Dithorium Dinitride Selenide, Th_2N_2Se

Th_2N_2Se was obtained by the reaction of cold pressed mixtures of ThSe and ThN in proper amounts at 1500 to 1700°C under 1 atm of nitrogen for $^1/_2$ to 2 h in a tungsten crucible [1]. Th_2N_2Se was also prepared by the reaction of ThN with pure selenium in stoichiometric amounts sealed in an evacuated silica tube and heated to 1000°C for 30 days [1], also cited in [2].

Th_2N_2Se is hexagonal (Ce_2O_2S type) with $Z = 1$; the space group is $P\bar{3}m1$-D_{3d}^3 (No. 164). The lattice parameters, obtained from X-ray diffraction, are a = 4.0287 ± 0.0002 Å, c = 7.156 ± 0.001 Å [1], see also [2, 3]. The X-ray density was calculated from these data to be 9.43 g/cm^3. The measured pycnometric density is 7.9 g/cm^3 [1]. The atomic positions due to the space group $P\bar{3}m1$ are 2 Th in ($^1/_3$, $^2/_3$, u_1), 2 N in ($^1/_3$, $^2/_3$, u_2), 1 Se in (0, 0, 0), with parameters $u_1 = 0.293$ (calculated from the observed intensities) and $u_2 = 0.628$ (if the nitrogen atoms are placed equidistantly from the four thorium atoms). The interatomic distances are Th-4 N = 2.39 Å, Th-3 Se = 3.13 Å [1], see also [3].

References for 11.3:

[1] R. Benz, W. H. Zachariasen (Acta Cryst. B **25** II [1969] 294/6). — [2] R. M. Dell, N. J. Bridger (MTP [Med. Tech. Publ. Co.] Intern. Rev. Sci. Inorg. Chem. Ser. One **7** [1972] 211/74). — [3] D. J. Lam, J. B. Darby Jr., M. V. Nevitt (Actinides Electron. Struct. Relat. Prop. **2** [1974] 119/84).

11.4   Compounds of Thorium with Selenium Oxoacids

David Brown
Chemistry Division, A.E.R.E.
Harwell, Oxon, England

11.4.1   Selenites

Early work on diselenite hydrates, $Th(SeO_3)_2 \cdot n\,H_2O$ (n = 8 and 1) and acid selenites such as 2 $Th(SeO_3)_2 \cdot 3\,H_2SeO_3 \cdot 13\,H_2O$ and $Th(SeO_3)_2 \cdot 3\,H_2SeO_3 \cdot 5\,H_2O$ is reviewed in "Thorium" 1955, p. 294. There has been little further work on Th^{IV} selenites since the publication of that review, and although solid diselenites have been precipitated by addition of selenious acid to aqueous media containing $ThCl_4$ or $Th(NO_3)_4$ the interest was analytical and the degree of hydration of the precipitates was not established [1 to 3], see also [4].

The solubility product $[Th^{4+}][SeO_3^{2-}]^2$ is determined to be 1.35×10^{-20}, from measurements of the solubility of $Th(SO_3)_2$ in aqueous HNO_3 and H_2SO_4 at 20°C [5].

References for 11.4.1:

[1] G. S. Deshmukh, L. K. Swamy (Anal. Chem. **24** [1952] 218/8). − [2] G. S. Deshmukh, V. D. Anand, P. Srinivasa (J. Sci. Ind. Res. [India] B **20** [1961] 356/71). − [3] V. D. Anand, B. Srinivasa Achar, G. S. Deshmukh (Z. Anorg. Allgem. Chem. **315** [1962] 309/14). − [4] Jen-Yin-Yen, Kuang-Hua Djao, Feng-Chiao Hsiao (Beijing Daxue Xuebao Ziran Kexueban **4** [1958] 195/9). − [5] E. I. Krylov, V. G. Chukhlantsev (Zh. Analit. Khim. **12** [1957] 451/6; J. Anal. Chem. [USSR] **12** [1957] 469/73).

11.4.2 Selenates

There appear to have been no publications on the preparation of thorium selenates since they were reviewed in "Thorium" 1955, p. 294, which contains information on the preparation and properties of $Th(SeO_4)_2 \cdot 8 H_2O$.

The estimated enthalpy of formation of $Th(SeO_4)_2$ is −458.8 kcal/mol [1].

Reference for 11.4.2:

[1] V. M. Amosov, V. E. Plyushchev (Izv. Vysshikh Uchebn. Zavedenii Khim. Khim. Tekhnol. **11** [1968] 1128/34).

12   Compounds of Thorium and Tellurium

Horst Wedemeyer
Kernforschungszentrum Karlsruhe
Institut für Material- und Festkörperforschung
Karlsruhe, Federal Republic of Germany

12.1   Binary Thorium Tellurides

12.1.1   The Th-Te System

The binary compounds Th_3Te [1], ThTe [2, 3], Th_2Te_3 (or Th_7Te_{12}) [4], $ThTe_2$ [2, 4], $ThTe_{2.66}$ [2], and $ThTe_3$ [4] have been claimed to be prepared by different authors, but only the compounds ThTe and $ThTe_2$ have been confirmed to exist in the system thorium-tellurium.

The compound Th_3Te, prepared by hydrogen reduction of $ThTeO_4 \cdot 8\,H_2O$ at 400°C [1], see also "Thorium" 1955, p. 295, could not be confirmed [2, 4]. No intermediate phases or eutectic compositions were reported to exist between thorium and ThTe. The cubic ThTe (CsCl type), black in color, has a melting point of 1680°C under flowing high-purity inert gas (1 to 3 atm) [5] or decomposes below 1000°C if heated in vacuum [2]. Th_2Te_3, black in color, is reported to crystallize in the hexagonal system. The detailed structure is not confirmed by X-ray measurements [4]. It is thought that the structure may belong to the Th_7X_{12} type [4, 6]. The compound degraded on heating in vacuum to 1050°C [4]. $ThTe_2$, black in color, was interpreted from the complex X-ray diffraction pattern to crystallize in the hexagonal system [4]. This compound decomposes in vacuum above 500°C, eventually forming the elements [2] and is reported to be stable up to 950°C [7]. The most tellurium-rich compound was found by tensimetric degradation to be $ThTe_{2.66}$ (Th_3Te_8), but the complex X-ray diffraction pattern could not be interpreted. The compound degraded readily when heated in vacuum [2]. $ThTe_3$, prepared by synthesis of the elements, seems to be monoclinic (isostructural with $ZrSe_3$) with a pronounced layer structure. This compound was slowly degraded to $ThTe_2$ when heated up in vacuum to 600°C and to $ThTe_{1.95}$ at 950°C [7].

It should be noted that all references to the composition $ThTe_{2.5}$ in [7] should be read $ThTe_3$ [4].

12.1.2   Thorium Monotelluride, ThTe

12.1.2.1   Formation and Preparation

Thorium monotelluride, ThTe, was prepared for the first time in 1954. The reaction was carried out with stoichiometric amounts of thorium and tellurium placed in evacuated (10^{-5} Torr) and sealed quartz tubes. The reaction mixture was slowly heated to 400°C, allowing a fair quantity of the tellurium to be absorbed by thorium. The strongly exothermic reaction started at 400°C. The temperature was then cautiously raised to about 600°C. This annealing was carried out with the sealed reaction tube placed in a vacuum tube (10^{-5} Torr) to prevent diffusion of oxygen through the silica walls, since this could oxidize the ThTe already formed to ThOTe or ThO_2 (main impurity of oxygen found 0.5%) [2]. To prevent the initial violent reaction a special apparatus was used as shown in **Fig. 36**a. With this method the thorium and tellurium were placed in separate limbs of the H-shaped apparatus. The apparatus was evacuated and sealed off at the constriction B, placed in an oven, and heated to 600°C to give a vapor pressure of tellurium of a few Torr. After a few hours, when the reaction was carried out, the thorium limb was sealed and drawn off at the constriction A. After the annealing step,

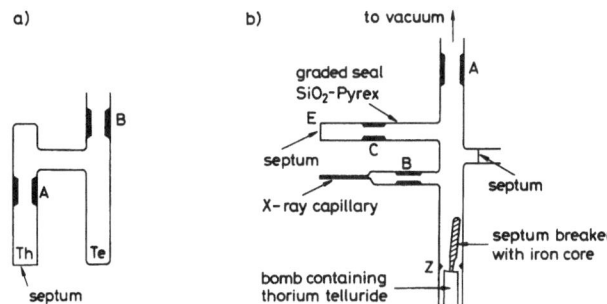

Fig. 36a. Scheme of apparatus for the preparation of thorium telluride compounds [2].

Fig. 36b. Special apparatus for the handling of thorium telluride compounds (schematic) [2].

the reaction tube was connected with a special apparatus, as shown in Fig. 36b, which allows one to open the tube under vacuum conditions and to separate samples for further inspections [2].

ThTe was also obtained as a black pyrophoric powder by heating the elements in quartz ampules at 900 °C [3]. There remains some doubt whether the ThTe prepared is an oxygen stabilized phase or not [4, 8]. It should be noted that the synthesis from the elements was reported to be unsuccessful after 2 weeks heating at 800 °C, then grinding under inert conditions, sealing and re-evacuating, and again heating for 7 days to 800 °C. No more than a small portion of the sample had reacted to form the monotelluride [4].

ThTe was also prepared by the reaction of thorium metal powder and zinc telluride, ZnTe, in stoichiometric amounts. The reaction was carried out at 700 °C leading to a highly pyrophoric powder, which had to be handled under inert conditions in glove boxes. The presence of ThTe was confirmed by X-ray diffraction but the pattern indicated the presence also of other telluride compounds. The oxygen content was measured to be about 0.71% [9].

The enthalpy of formation of ThTe was estimated to be $\Delta H_f^\circ(\text{ThTe}, c, 298) = -58 \pm 10$ kcal/mol [10], see also [6].

12.1.2.2 Crystallographic Properties

ThTe is body-centered cubic (CsCl type) with $Z = 1$; the space group is $Pm3m\text{-}O_h^1$ (No. 221) [2]. The measured lattice constants are summarized in Table 56.

Table 56
Measured Lattice Constants of ThTe.

a	Ref.
3.819 ± 0.002 kX	[2], see also [11]
or 3.826 Å	[12], from [2], see also [8]
or 3.827 ± 0.002 Å	[4], from [2], see also [6, 13]
3.828 Å	[3], see also [6]
3.832 Å	[5]
3.824 Å	[14], see also [6]

References for 12.1 on p. 114

The X-ray density (in g/cm^3) was calculated to be 10.63 [2], see also [13], 10.65 [3], or 10.61 [5]. From the X-ray data the Th-Te distance was calculated to be Th-8 Te = 3.31 kX [2], or 3.31 Å [3]. The cation radius (Te^{2-}), from cation-anion separation experiments, is 2.135 Å [15].

12.1.2.3 Thermal Properties

ThTe decomposes below 1000°C, if heated in vacuum [2]. A melting point of 1680°C was observed when ThTe was heated in flowing high-purity inert gas at 1 to 3 atm [5].

The entropy of ThTe was estimated to be $S°(ThTe, c, 298) = 24 \pm 3$ cal · mol^{-1} · K^{-1}, or 23 ± 3 cal · mol^{-1} · K^{-1} [10], see also [6, 17]. Estimated values for the thermodynamic functions of gaseous ThTe are reported in [6]. The dissociation energy is assumed to be $D_0^\circ = 95 \pm 10$ kcal/mol and $D_{298}^\circ = 96 \pm 10$ kcal/mol. From this, the enthalpy of formation follows to be $\Delta H_{f,298}^\circ = 97 \pm 10$ kcal/mol [6]. Further detailed data of the thermodynamic functions of gaseous ThTe are given in [6], see Table 57.

12.1.2.4 Electrical Properties

The electric conductance of ThTe shows a metallic character. A value of the room temperature resistivity, measured on pressed powders (25 000 psi), is 0.02 Ω · cm [18].

12.1.2.5 Chemical Behavior

ThTe, black in color, is degraded below 1000°C when heated in vacuum [2]. Otherwise, ThTe is reported to be stable up to 950°C [7]. A melting point of 1680°C, measured in flowing inert gas at 1 to 3 atm, is given for ThTe [5].

ThTe powder is highly reactive, sometimes pyrophoric in air [3, 9], see also [7]; the oxidation products are ThO_2 and TeO_2 [7], see also [2]. It reacts with chlorine below 300°C [7]. The alkali metals reduce all tellurides to the metal, regardless of composition [7].

The tellurides were leached out from thorium- or thorium dioxide-containing samples in HNO_3 at about 30°C. The samples were first covered with water before addition of HNO_3 to smooth the extremely vigorous reaction [2], see also [7]. All chalcogenides are attacked by boiling H_2SO_4 [7]. All tellurides react completely with concentrated HCl, usually with no evolution of gas [7].

12.1.3 Dithorium Tritelluride, Th_2Te_3 (or Heptathorium Dodecatelluride, Th_7Te_{12})

12.1.3.1 Formation and Preparation

A compound, Th_2Te_3, was prepared by the reaction of thorium powder and tellurium. The reaction was carried out in an evacuated and sealed quartz tube at 800°C for two weeks. The silica tube was then opened in dry argon, the reaction product thoroughly ground up, the tube was re-evacuated and sealed, and again heated to 800°C for another 7 days [4].

A value of the enthalpy of formation for $ThTe_{1.50}$ is given as $\Delta H_{f,298}^\circ = -75$ kcal/mol [7], see also [17], or -80 ± 12 kcal/mol [6].

Table 57
Thermodynamic Functions of Gaseous ThTe(g) [6]. See note in Table 9 on p. 15.

T in K	C°p in cal·mol⁻¹·K⁻¹	S° in cal·mol⁻¹·K⁻¹	$-(G°_T - H°_{298})/T$ in cal·mol⁻¹·K⁻¹	$H°_T - H°_{298}$ in cal/mol	$\Delta H°_f$ in cal/mol	$\Delta G°_f$ in cal/mol	log K$_p$
298	8.761	65.090	65.090	0	97000	84816	−62.171
300	8.763	65.144	65.090	16	96993	84831	−61.798
400	8.888	67.680	65.435	898	96591	80836	−44.166
500	9.119	69.684	66.091	1797	96172	76945	−33.632
600	9.538	71.382	66.835	2728	95755	73139	−26.641
700	10.112	72.894	67.594	3710	95349	69402	−21.668
800	10.736	74.286	68.345	4753	90584	66181	−18.080
900	11.303	75.585	69.078	5856	90058	63169	−15.339
1000	11.737	76.799	69.790	7010	89617	60198	−13.156
1100	12.011	77.932	70.479	8198	89223	57277	−11.380
1200	12.133	78.983	71.145	9406	88837	54389	−9.905
1300	12.131	79.955	71.785	10620	75542	51735	−8.697
1400	12.039	80.851	72.401	11829	75361	49910	−7.791
1500	11.888	81.677	72.993	13026	75151	48099	−7.008
1600	11.704	82.438	73.559	14206	74906	46304	−6.325
1700	11.506	83.142	74.103	15366	73808	44559	−5.728
1800	11.306	83.794	74.623	16507	73548	42846	−5.202
1900	11.112	84.400	75.122	17628	73239	41148	−4.733
2000	10.929	84.965	75.600	18730	72883	39468	−4.313
2100	10.759	85.494	76.059	19814	69077	37935	−3.948
2200	10.604	85.991	76.499	20882	68519	36465	−3.622
2300	10.463	86.459	76.922	21935	67946	35021	−3.328
2400	10.335	86.902	77.329	22975	67360	33602	−3.060
2500	10.219	87.321	77.720	24003	66761	32207	−2.816
2600	10.115	87.720	78.097	25019	66149	30837	−2.592
2700	10.022	88.100	78.461	26026	65528	29490	−2.387
2800	9.937	88.463	78.811	27024	64897	28167	−2.199
2900	9.861	88.810	79.150	28014	64257	26867	−2.025
3000	9.793	89.143	79.478	28996	63608	25588	−1.864

References for 12.1 on p. 114

12.1.3.2 Crystallographic Properties

The compound Th_2Te_3 is reported to crystallize in the hexagonal system with the lattice constants a = 12.49 Å and c = 4.354 Å [4], see also [6, 8, 13]. The unit cell is similar to those reported for Th_7S_{12} (see Section 10.1.4.2, p. 29) and Th_7Se_{12} (see Section 11.1.4.2, p. 95), and the observed X-ray reflections are consistent with the same space group, $P6_3/m-C_{6h}^2$ (No. 176), also cited in [6, 13]. From this, the composition suggests a defect cation structure, which is reasonable, since the thorium positions are fully occupied in the Th_7X_{12} structure [4]. The density of the compound was measured to be 8.6 g/cm^3 [4], also cited in [6]. The calculated density for $Th_7Te_{10.5}$ would be 8.35 g/cm^3. This discrepancy is suggested to be due to the presence of some ThO_2. The stoichiometric composition Th_7Te_{12} would lead to a density of 8.89 g/cm^3, a value too high [4], see also [6]. A range of nonstoichiometry has been suggested [4].

12.1.3.3 Thermal Properties

The compound Th_2Te_3 was degraded on heating in vacuum to 1050°C for more than one day [4].

Estimated values for the standard entropy are $S°(ThTe_{1.5}, c, 298) = 29.3$ cal $\cdot$ mol^{-1} $\cdot$ K^{-1} [16], see also [17] or $S°(ThTe_{1.71}, c, 298) = 30 \pm 3$ cal $\cdot$ mol^{-1} $\cdot$ K^{-1} [6].

12.1.3.4 Electrical Properties

The electric conductance of Th_2Te_3 has a metallic character. A value of the room temperature resistivity, measured on pressed powders (25000 psi), is 0.015 $\Omega \cdot$ cm [18].

12.1.3.5 Chemical Behavior

The compound Th_2Te_3, black in color, degraded on heating in vacuum to 1050°C for more than one day [4].

All chalcogenides are oxidized when heated in air or oxygen to form the metal oxide (ThO_2) and the oxide of the chalcogen (TeO_2) [7]. All chalcogenides react with chlorine below 300°C [7]. The alkali metals reduce all tellurides to the metal, regardless of composition [7].

All chalcogenides react with cold concentrated HNO_3 and boiling H_2SO_4 [7]. All chalcogenides react completely with concentrated HCl, usually with no evolution of gas [7].

12.1.4 Thorium Ditelluride, ThTe$_2$

12.1.4.1 Formation and Preparation

Thorium ditelluride was prepared for the first time in 1954. The reaction was carried out with stoichiometric amounts of thorium and tellurium placed in evacuated (10^{-5} Torr) and sealed quartz tubes. The reaction mixture was slowly heated to 400°C, allowing a fair quantity of the tellurium to be absorbed by thorium. The strongly exothermic reaction started at 400°C; the temperature was then cautiously raised to about 600°C. $ThTe_2$ was obtained as a black powder (main impurity of oxygen about 0.5%) [2]. For further details, see Section 12.1.2.1, p. 106.

ThTe$_2$ was also obtained by degradation of ThTe$_{2.66}$ in a special H-shaped apparatus (shown in Fig. 36a, p. 107) after heating at 800°C for one week in vacuum. The chemical analysis resulted in a Te/Th ratio of 1.94 [2]. The degradation of a compound with the gross composition of about ThTe$_3$ led to the formation of ThTe$_{2.0}$ (Th = 47.4%, Te = 52.4%) when heated in vacuum at 900°C for several hours [4]. Further heating in vacuum to 800°C for a few weeks led to a sample with a composition close to ThTe$_{1.9}$, which is reported to have a different phase structure than the sample with composition ThTe$_{2.0}$ [4]. Degradation at 950°C in vacuum led to the formation of ThTe$_{1.95}$ [7].

A value of the enthalpy of formation was estimated by analogy with values for related compounds to be $\Delta H°(ThTe_2, c, 298) = -87 \pm 12$ kcal/mol [10], see also [6].

12.1.4.2 Crystallographic Properties

ThTe$_2$ gave only poor [2] or complex [4] X-ray diffraction patterns, which were dissimilar to those of ThSe$_2$ [2]. The X-ray reflections were indexed as hexagonal with the lattice parameters a = 8.49, c = 9.01 Å [4], also cited in [6, 13], but this might only be a pseudo-cell [4].

A ThTe$_{1.9}$ sample, obtained from a ThTe$_2$ sample heated to 800°C in vacuum for a few weeks, was interpreted as hexagonal with a = 12.33 Å (about the order of magnitude required for the Th$_7$X$_{12}$ type) but a c axis of the required magnitude could not be found; most of the reflections were accounted for if c = 13.8 Å [4], also cited in [6].

12.1.4.3 Thermal Properties

ThTe$_2$ decomposes in vacuum above 500°C, eventually forming the elements [2], see also [7, 8]. Otherwise, ThTe$_2$ was observed to be stable up to 950°C [7], see also [6, 8].

Estimated values for the standard entropy, S°, are $S°(ThTe_2, c, 298) = 32 \pm 3$ cal · mol^{-1} · K^{-1} [10], see also [6], or 34 ± 3 cal · mol^{-1} · K^{-1} [16], see also [6, 17].

Estimated values for the thermodynamic functions of gaseous ThTe$_2$ are reported in [6]. The atomization energy is assumed to be $D°_{at, 298} = 175 \pm 20$ kcal/mol. From this, the enthalpy of formation is calculated to be $\Delta H°_{f, 298} = 69 \pm 20$ kcal/mol [6]. Further detailed data of the thermodynamic functions of gaseous ThTe$_2$ are given in [6], see Table 58, p. 112.

12.1.4.4 Electrical Properties

The electric conductance of the compound ThTe$_2$ has a metallic character. A value of the room temperature resistivity, measured on pressed powders (25000 psi), is 0.02 Ω · cm [18].

12.1.4.5 Chemical Behavior

ThTe$_2$, black in color, decomposes in vacuum above 500°C [2] or 950°C [7], eventually forming the elements.

ThTe$_2$ was degraded to ThTe$_{1.9}$ when heated in highly purified hydrogen to 950°C for 3 h [7]. All chalcogenides are oxidized when heated up in air or oxygen to form the metal oxide (ThO$_2$) and the oxide of the chalcogen (TeO$_2$) [7], see also [2]. All chalcogenides react with chlorine below 300°C [7]. The alkali metals reduce all tellurides to the metal, regardless of composition [7].

References for 12.1 on p. 114

Table 58
Thermodynamic Functions of Gaseous ThTe$_2$(g) [6]. See note in Table 9 on p. 15.

T in K	C_p° in cal·mol^{-1}·K^{-1}	S° in cal·mol^{-1}·K^{-1}	$-(G_T^\circ - H_{298}^\circ)/T$ in cal·mol^{-1}·K^{-1}	$H_T^\circ - H_{298}^\circ$ in cal/mol	ΔH_f° in cal/mol	ΔG_f° in cal/mol	log K_p
298	13.472	81.600	81.600	0	69000	55311	−40.544
300	13.477	81.683	81.600	25	68991	55407	−40.364
400	13.661	85.588	82.131	1383	68446	50960	−27.843
500	13.749	88.646	83.140	2753	67867	46654	−20.392
600	13.797	91.158	84.273	4131	67256	42469	−15.469
700	13.827	93.287	85.412	5512	66588	38389	−11.985
800	13.846	95.135	86.515	6896	57102	35327	−9.651
900	13.860	96.766	87.565	8281	55995	32684	−7.937
1000	13.869	98.227	88.559	9668	54978	30135	−6.586
1100	13.876	99.549	89.499	11055	54005	27699	−5.503
1200	13.882	100.757	90.388	12443	53029	25350	−4.617
1300	13.886	101.868	91.229	13831	26241	23484	−3.948
1400	13.889	102.897	92.026	15220	25711	23292	−3.636
1500	13.892	103.856	92.783	16609	25166	23139	−3.371
1600	13.894	104.752	93.503	17998	24603	23022	−3.145
1700	13.896	105.595	94.190	19388	23208	22975	−2.954
1800	13.898	106.389	94.846	20778	22672	22977	−2.790
1900	13.899	107.141	95.473	22167	22105	23008	−2.647
2000	13.900	107.853	96.075	23557	21512	23072	−2.521
2100	13.901	108.532	96.652	24947	17487	23293	−2.424
2200	13.902	109.178	97.207	26338	16726	23586	−2.343
2300	13.903	109.796	97.741	27728	15964	23916	−2.272
2400	13.903	110.388	98.256	29118	15202	24278	−2.211
2500	13.904	110.956	98.752	30509	14439	24672	−2.157
2600	13.904	111.501	99.232	31899	13673	25096	−2.109
2700	13.905	112.026	99.696	33289	12907	25548	−2.068
2800	13.905	112.531	100.146	34680	12140	26031	−2.032
2900	13.905	113.019	100.581	36070	11370	26541	−2.000
3000	13.906	113.491	101.004	37461	10599	27077	−1.973

The tellurides were leached out in HNO_3 from thorium- and thorium oxide-containing samples at about 30°C. The samples were covered with water before addition of HNO_3 to smooth the extremely vigorous reaction [2], see also [7]. All chalcogenides are attacked by boiling H_2SO_4 [7]. They react completely with concentrated HCl, usually with no evolution of gas [7]. $ThTe_2$ is reported to react with TiS_2 to form ThS_2 and $TiTe_2$ [7]. $ThTe_2$ appeared to react to a very limited extent with n-butyl alcohol, forming $Th(C_4H_9O)_4$ [7].

12.1.5 Thorium Tritelluride, $ThTe_3$

12.1.5.1 Formation and Preparation

From tensimetric degradation studies, starting with a compound of gross composition $ThTe_3$, the upper limiting composition of the Th-Te system was claimed to be $ThTe_{2.66}$ (Th_3Te_8) with a complex X-ray diffraction pattern. It was readily degraded to $ThTe_2$ when heated in vacuum [2], cited also in [6, 13]. The preparation of a compound $ThTe_{3.05}$ and a second compound $ThTe_{2.95}$ (with sharp X-ray patterns) led to the conclusion that the highest telluride in the system is $ThTe_3$ [4], also cited in [6]. It should be noted that the composition referred to as Th_2Te_5 in [7] should be read $ThTe_3$ [4].

$ThTe_3$ was obtained by reaction of the elements using excess tellurium, placed in an evacuated and sealed quartz tube, at a reaction temperature of 600°C for 2 to 3 weeks to ensure complete reaction and reasonable crystal growth. The excess tellurium was then sublimed off by heating the reaction product to 450°C. Two syntheses were carried out resulting in $ThTe_{3.05}$ and $ThTe_{2.95}$, respectively. Both materials gave sharp X-ray patterns [4].

The enthalpy of formation of $ThTe_3$ was estimated to be $\Delta H_f^{\circ}(ThTe_3, c, 298) = -90 \pm 15$ kcal/mol [6].

12.1.5.2 Crystallographic Properties

$ThTe_3$ seems to crystallize in the monoclinic system (isostructural with $ZrSe_3$), as observed from the sample with composition $ThTe_{2.95}$, with the lattice constants a = 6.14 Å, b = 4.31 Å, c = 10.44 Å, β = 98.4° [4], see also [6].

The X-ray density, calculated from the monoclinic cell, is 7.44 g/cm³ as compared to the measured density of 7.40 g/cm³ [4]. $ThTe_3$ shows a pronounced layer structure with an (001) spacing of 10.36 Å [4].

12.1.5.3 Thermal Properties

$ThTe_3$ was degraded slowly to $ThTe_2$ when heated up in vacuum to 600°C [4] and to $ThTe_{1.95}$ at 950°C [7].

The standard entropy, S°, of $ThTe_3$ was estimated to be $S^{\circ}(ThTe_3, c, 298) = 43.5 \pm 4$ cal · $mol^{-1} \cdot K^{-1}$ [16], see also [6].

12.1.5.4 Electrical Properties

The electric conductance of the compound $ThTe_3$ shows a metallic (n-type) character. A value of the room temperature resistivity, measured at pressed powders (25000 psi), is 4.0 Ω · cm [18], see also [4].

12.1.5.5 Chemical Behavior

ThTe$_3$, black in color, was degraded in vacuum at 600°C to give ThTe$_2$ [4] and at 950°C to ThTe$_{1.95}$ [7].

ThTe$_3$ reacts with air forming ThO$_2$ and Te [4]. All Th chalcogenides are oxidized when heated in air or oxygen to form the metal oxide (ThO$_2$) and the oxide of the chalcogen (TeO$_2$). All Th chalcogenides react with chlorine below 300°C. The alkali metals reduce all Th tellurides to the metal, regardless of composition [7].

All Th chalcogenides react with concentrated HNO$_3$; the reaction is vigorous. All Th chalcogenides are attacked by boiling H$_2$SO$_4$. The Th tellurides react completely with concentrated HCl, usually with no evolution of gas. ThTe$_3$ (reported as Th$_2$Te$_5$) reacts also with 50 vol% H$_2$O$_2$ [7].

References for 12.1:

[1] E. Montiguie (Bull. Soc. Chim. France **1947** 748/9). — [2] R. W. M. D'Eye, P. G. Sellman (J. Chem. Soc. **1954** 3760/6). — [3] F. Ferro (Atti Accad. Nazl. Lincei Rend. Classe Sci. Fis. Mat. Nat. [8] **18** [1955] 641/4). — [4] J. Graham, F. K. McTaggart (Australian J. Chem. **13** [1960] 67/73). — [5] J. H. Handwerk, O. L. Kruger (Nucl. Eng. Design **17** [1971] 397/408).

[6] F. Grønvold, J. Drowart, E. F. Westrum Jr. (in: F. L. Oetting, The Chemical Thermodynamics of Actinide Elements and Compounds, Pt. 4, IAEA, Vienna 1984, pp. 27/32). — [7] J. Bear, F. K. McTaggart (Australian J. Chem. **11** [1958] 458/70). — [8] R. M. Dell, N. J. Bridger (MTP [Med. Tech. Publ. Co.] Intern. Rev. Sci. Inorg. Chem. Ser. One **7** [1972] 211/74). — [9] G. H. B. Lovell, D. R. Perels, E. J. Britz (J. Nucl. Mater. **39** [1971] 303/10). — [10] K. C. Mills (Thermodynamic Data for Inorganic Sulphides Selenides and Tellurides, Butterworth, London 1974).

[11] D. J. Lam, J. B. Darby Jr., M. V. Nevitt (Actinides Electron. Struct. Relat. Prop. **2** [1974] 119/84). — [12] A. A. Bauer, F. A. Rough (Progr. Nucl. Energy Ser. V **2** [1959] 612/20). — [13] K. Girgis (At. Energy Rev. Spec. Issue No. 5 [1975] 191/238). — [14] M. Haessler, C. H. De Novion, D. Damien (Plutonium 1975 Other Actinides Proc. 5th Intern. Conf., Baden-Baden, FRG, 1975 [1976], pp. 649/57). — [15] M. Allbutt, R. M. Dell (J. Inorg. Nucl. Chem. **30** [1968] 705/10).

[16] E. F. Westrum Jr., F. G. Grønvold (Thermodyn. Nucl. Mater. Proc. Symp., Vienna 1962, pp. 23/36; SM-26/30 [1963] 1/15). — [17] M. H. Rand (At. Energy Rev. Spec. Issue No. 5 [1975] 7/86). — [18] F. K. McTaggart (Australian J. Chem. **11** [1958] 471/80).

12.2 Compounds of Thorium with Tellurium and Oxygen

In this chapter, only the one existing thorium oxide telluride, ThOTe, is treated. Thorium tellurites and tellurates are described in Chapter 12.4, pp. 116.

12.2.1 Thorium Oxide Telluride, ThOTe

12.2.1.1 Preparation

ThOTe was prepared by the reaction of a mixture of stoichiometric amounts of thorium, tellurium, and ThO$_2$; the ThO$_2$ was prepared from thorium oxalate to obtain a ThO$_2$ powder with small particle size. The reaction was carried out in evacuated and sealed silica tubes at 1100°C over night [1], see also [2]. ThOTe was also prepared by careful oxidation of ThTe (see p. 106) [2].

Single crystals of ThOTe were obtained from the elements (Th, Te) mixed with a third element (Si, Ge, P, As, Sb, or Bi) by a transport method with bromine or iodine as transporting agent (3 to 6 mg/cm³). The reactants are placed in uncoated quartz ampules heated to 900 to 1050°C at one end and 50 to 100°C cooler at the other, for one week. With this method the SiO_2 is reduced to elemental silicon (which may react with the third element to form SiP, SiAs, ...) and the oxygen is used for the formation of ThOTe. The crystals grew as thin plate-lets, 1 to 3 mm edge length, which were metallic gray in color (black when powdered) [3], also cited in [4].

12.2.1.2 Crystallographic Properties

ThOTe is tetragonal (PbFCl type) with $Z = 2$; the space group is $P4/nmm-D_{4h}^7$ (No. 129) [1], see also [2, 4, 5]. The lattice parameters, measured by X-ray diffraction, are a = 4.112 ± 0.005 kX, c = 9.544 (7.544 [4]) kX [1], see also [5], or a = 4.12 Å, c = 7.56 Å [4] from [1]. From this the X-ray density was calculated to be 9.72 g/cm³ [1], see also [2, 5]. The atomic positions for the PbFCl structure are 2 Th in $(^1/_2, 0, x)$, $(0, ^1/_2, \bar{x})$; 2 O in $(0, 0, 0)$, $(^1/_2, ^1/_2, 0)$; 2 Te in $(^1/_2, 0, z)$, $(0, ^1/_2, \bar{z})$; with parameters $x = 0.18$ and $z = 0.65$ [1]. The interatomic distances are Th-4 O = 2.46 kX, Th-4 Te = 3.18 kX, Th-Te = 3.18 kX [1], see also [2]. A coordination polyhedron for the isostructural ThOSe is shown in Fig. 35, p. 102.

12.2.1.3 Thermal Properties

ThOTe was degraded when heated in vacuum to ThO_2, Th, and Te [1], see also [6].

References for 12.2:

[1] R. W. M. D'Eye, P. G. Sellman (J. Chem. Soc. **1954** 3760/6). − [2] F. Ferro (Atti Accad. Nazl. Lincei Rend. Classe Sci. Fis. Mat. Nat. [8] **18** [1955] 641/4). − [3] H. U. Boelsterli, F. Hulliger (J. Mater. Sci. **3** [1968] 664/5). − [4] R. M. Dell, N. J. Bridger (MTP [Med. Tech. Publ. Co.] Intern. Rev. Sci. Inorg. Chem. Ser. One **7** [1972] 211/74). − [5] K. Girgis (At. Energy Rev. Spec. Issue No. 5 [1975] 191/238).

[6] J. Bear, F. K. McTaggart (Australian J. Chem. **11** [1958] 458/70).

12.3 Compounds of Thorium with Tellurium and Nitrogen

There is only one compound known within the system thorium-tellurium-nitrogen: Th_2N_2Te.

12.3.1 Dithorium Dinitride Telluride, Th_2N_2Te

12.3.1.1 Preparation

Th_2N_2Te was obtained by the reaction of cold pressed mixtures of ThTe and ThN in proper amounts at 1200°C in an atmosphere of nitrogen within 2 h. The compound was also prepared by the reaction of ThN with pure tellurium in stoichiometric amounts sealed in an evacuated silica tube and heated to 1000°C for one month [1].

12.3.1.2 Crystallographic Properties

Th_2N_2Te is body-centered tetragonal (U_2N_2Sb type) with Z = 2; the space group is I4/mmm-D_{4h}^{17} (No. 139) [1], see also [2]. The lattice parameters, measured from X-ray diffraction patterns, are a = 4.0939 ± 0.0004 Å, c = 13.014 ± 0.001 Å [1], also cited in [2]. From this, the X-ray density was calculated to be 9.44 g/cm^3 [1]. The atomic positions, with the origin chosen at a Te atom, are equivalent positions: (0,0,0; $^1/_2$,$^1/_2$,$^1/_2$) + 2 Te in (0,0,0), 4 Th in ± (0,0,u), 4 N in ($^1/_2$,0,$^1/_4$); (0,$^1/_2$,$^1/_4$), with u = 0.344 ± 0.003 [1], see also [2]. The interatomic distances are Th-4 N = 2.38 Å, Th-4 Te = 3.54 Å [1].

References for 12.3:

[1] R. Benz, W. H. Zachariasen (Acta Cryst. B **26** [1970] 823/7). − [2] D. J. Lam, J. B. Darby Jr., M. V. Nevitt (Actinides Electron. Struct. Relat. Prop. **2** [1974] 119/84).

12.4 Compounds of Thorium with Tellurium Oxoacids

David Brown
Chemistry Division, A.E.R.E.
Harwell, Oxon, England

12.4.1 Thorium Tellurites

Thorium tellurites have only recently been characterised and the only known compounds are $Th(TeO_3)_2$, $Ba_5Th(TeO_3)_7$ and $Ca_5Th(TeO_3)_7$.

12.4.1.1 Thorium Tellurite, $Th(TeO_3)_2$

$Th(TeO_3)_2$, obtained by heating together ThO_2 and TeO_2 at 650 to 700 °C in an atmosphere of argon [1, 2], is a white solid which is reported to possess orthorhombic symmetry with a = 5.89, b = 10.71, c = 12.10 Å (Z = 6) [1]. The product obtained by heating the component oxides in air at 600 °C [3] has the same composition and a similar X-ray powder pattern, based on a comparison of the published d values, but the unit cell deduced for this product is of cubic symmetry with a = 21.838(8) Å (Z = 64, D_m = 5.7 g · cm^{-3}). Additional crystallographic studies are required to ascertain the true symmetry of $Th(TeO_3)_2$.

Infrared spectra (1000 to 200 cm^{-1}) are illustrated in references [2, 3] and band positions are listed in [3] without assignment.

Thermal effects observed in argon at 880 °C (or at 897 °C [2]) and 1116 °C (or 1137 °C [2]) are attributed to decomposition (to ThO_2 and TeO_2) followed by melting of the resulting mixture [1, 2]. Loss of TeO_2 is observed in air at 877 °C. Partial conversion to ThO_2 and Te is observed above 447 °C in hydrogen [2].

References for 12.4.1.1:

[1] J. Wroblewska, J. Dobrowolski, M. Pages, W. Freundlich (Radiochem. Radioanal. Letters **39** [1979] 241/6). − [2] Yu. G. Marishov, V. B. Yadrintsev (Radiokhimiya **24** [1982] 256/7; Soviet Radiochem. **24** [1982] 215/6). − [3] I. L. Botto, E. J. Baran (Z. Anorg. Allgem. Chem. **484** [1982] 215/20).

12.4.1.2   Thorium Tellurito Complexes, $M_5^{II}Th(TeO_3)_7$

The complexes $Ba_5Th(TeO_3)_7$ and $Ca_5Th(TeO_3)_7$ have been identified during an investigation of the $MO-ThO_2-TeO_2$ (M = Ba, Ca) phase systems. They can be obtained pure by heating 5:1 mixtures of $MTeO_3$ and $Th(TeO_3)_2$ at 650°C for 24 h (M = Ba) or at 800°C for 4 h (M = Ca) [1]. The $Th(TeO_3)_2-BaTeO_3$ phase diagram is illustrated in **Fig. 37** where I indicates the $Ba_5Th(TeO_3)_7$ double salt. X-ray powder diffraction data for both complexes are given in [1]. They do not show the Ca and the Ba salts to be isostructural [1].

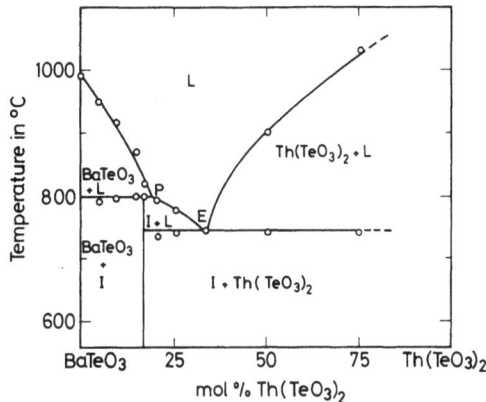

Fig. 37. The $Th(TeO_3)_2-BaTeO_3$ phase diagram. A peritectic point (= P) is at 800°C, a eutectic point (= E) at 742°C [1].

Reference for 12.4.1.2:

[1] J. Wroblewska, A. Erb, J. Dobrowolski, W. Freundlich (Ann. Chim. [Paris] [15] **4** [1979] 353/8).

12.4.2   Thorium Tellurates

No publications on the preparation of thorium tellurates seem to have appeared since the only known phases, $ThO(TeO_4) \cdot n H_2O$ (n = 8 and 4), were reviewed in "Thorium" 1955, p. 295.

The enthalpy of formation of $Th(TeO_4)_2$ has been estimated as -512.6 kcal/mol [1].

Reference for 12.4.2:

[1] V. M. Amosov, V. E. Plyushchev (Izv. Vysshikh Uchebn. Khim. Khim. Tekhnol. **11** [1968] 1128/34).

13   Compounds of Thorium and Boron

Horst Wedemeyer
Kernforschungszentrum Karlsruhe
Institut für Material- und Festkörperforschung
Karlsruhe, Federal Republic of Germany

13.1   Binary Thorium Borides

13.1.1   The Th-B System

The binary compounds ThB_4, ThB_6, ThB_{12}, and the "hectoboride" ThB_{66} are reported to exist in the thorium-boron system, based on X-ray diffraction measurements. The existence of the binary compounds ThB and a face-centered cubic ThB_2 (proposed by [1], see also [2 to 4]) could not be confirmed. The observed "ThB_2" was assumed to be a $Th(O, B)_2$ solid solution [5]. A further compound ThB_{18} was prepared from the elements, but its structure has not yet been determined [6].

The solubility of boron in thorium is reported to be probably small [7]. No intermediate phases were found to exist between β-Th and ThB_4, but a eutectic composition was observed with a eutectic temperature of 1450°C [8] or 1550°C [5], which is close to pure thorium [8]. The tetragonal (UB_4 type) thorium tetraboride, gray in color, has no [5] or a small homogeneity range [8]. The melting point of ThB_4 was measured to be about 2500°C [8,9]. A eutectic composition which melts at about 2400°C was found to exist between the phases ThB_4 and ThB_6 [8, 9]. The cubic (CaB_6 type) thorium hexaboride, red to red-violet in color, has a homogeneity range within the limits of $0 \leqslant x \leqslant 0.22$ according to $Th_{1-x}\square_xB_6$ [10]. Its melting point is reported to be about 2450°C [8] or 2190 to 2195°C [9, 11]. The lower melting points were assumed to be measured on samples containing free boron [8, 12]. The existence of a ThB_6-boron eutectic composition is reported with a melting point of about 2050°C [8]. A cubic (UB_{12} type) ThB_{12}, silvery in color, was prepared, which melts incongruently forming ThB_6 and a melt at high pressures and at fairly low temperature [13]. Furthermore, the existence of a cubic (YB_{66} type) thorium "hectoboride", ThB_{66}, is reported [14, 15].

A tentative phase diagram, based on the measured melting points and metallographic examinations [8], is given in **Fig. 38**.

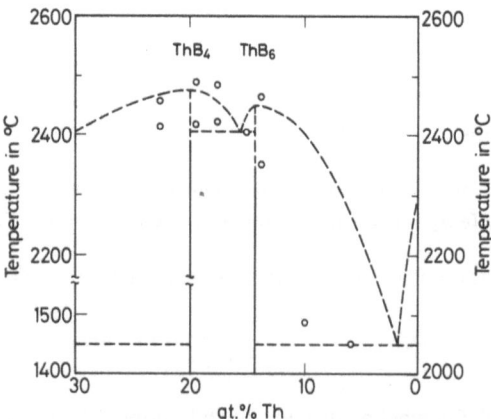

Fig. 38. Tentative phase diagram for the thorium-boron system [8].

13.1.2 Gaseous Thorium Monoboride

The existence of an ionic species ThB$^+$ was observed from effusion measurements at 2804 K using a Knudsen cell combined with a mass spectrometer. The measurements were performed on the gaseous equilibria over a thorium-boron alloy that contained a small amount of phosphorus. The dissociation energy D_0(ThB) = 69.9 $\pm$ 8 kcal/mol and the heat of formation ΔH_f°(ThB, g, 298) = 199.7 $\pm$ 12.5 kcal/mol were obtained from the experimentally determined enthalpies combined with published thermodynamic data. A vibrational frequency of 430 cm^{-1} was estimated [16].

13.1.3 Thorium Tetraboride, ThB$_4$

13.1.3.1 Formation and Preparation

The formation of thorium borides (ThB$_4$; ThB$_6$, p. 127) was observed in 1905 for the first time when ThO$_2$ was reduced with elemental boron in an electric furnace [17], see also [18], and in 1909 by direct synthesis from the elements [19]. Detailed information of the older literature is given in "Thorium" 1955, pp. 295/6.

Synthesis from the Elements

The reaction of thorium (powder) and boron (powder) starts at 850°C [6] or at 950°C under vacuum conditions when a sharp rise in temperature is observed [5]. This reaction was carried out in a high-frequency furnace at temperatures of max. 2050°C with the reactants mixed in stoichiometric amounts and placed in a molybdenum crucible. The reaction product was not sintered or shrunk together under those conditions [5], see also [12]. Different reaction temperatures and furnaces have been used for the preparation of ThB$_4$. At 1500°C a powdered reaction product was obtained which contained clusters of single crystals of a few tenths of a millimeter in size [20], see also [21]. Reaction took place at 1400°C [22, 23] or at 1600°C within 10 h with the reactants pressed into pellets, placed in a tungsten crucible, and fired in a vacuum furnace at 10^{-4} to 10^{-5} Torr [8]. The reaction started at 850°C with the reactants pressed into pellets and heated in a high-frequency induction furnace under a vacuum of 10^{-6} Torr [6]. ThB$_4$ was also obtained by arc-melting of the elements [24], see also [9, 12, 25].

The reaction of thorium powder and boron powder, pressed into pellets in stoichiometric amounts, in a stream of argon started at 1000°C. The reaction was carried out in a vertical furnace constructed from silica (see **Fig. 39**, p. 120). The reaction product contained 1 to 5% ThO$_2$ and similar amounts of ThB$_6$ [26], see also [12, 27].

Reduction of ThO$_2$ with Boron (and Carbon and/or B$_4$C)

Mixed powders of thorium dioxide, boron, and carbon, pressed into pellets (5 tsi) and placed in a graphite crucible, were heated in an induction furnace in an atmosphere of hydrogen at 1500°C for 2 h or at 1800°C for 1 h to obtain ThB$_4$. There were strong indications, especially for the reaction at 1800°C, that boron rather than carbon acted as the reducing agent with liberation of B$_2$O$_3$ [28], see also [29]. ThB$_4$ was also readily formed at 1500°C and 1800°C when the initial powder mixture contained no carbon. In this case the reaction product contained 1 to 2% carbon picked up from the graphite crucible. ThB$_4$ was equally well obtained if ZrB$_2$ crucibles were used [28], see also [29], also referred in [12]. The thermal reduction of thorium oxide with boron carbide, B$_4$C (see "Boron" Suppl. Vol. 2, 1981, p. 204), and carbon (lamp-black) — the powders were mixed for 16 h, passed three times through a 20 to 40 mesh

References for 13.1 on pp. 135/8

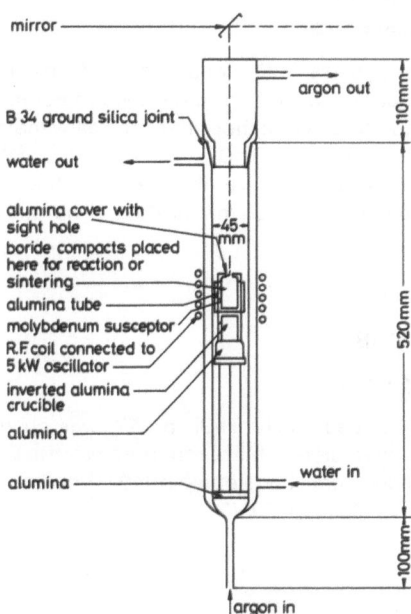

Fig. 39. Vertical furnace constructed from silica for the preparation of ThB$_4$ [27].

sieve, and pressed into pellets (ca. 1 t/cm^2) — at temperatures in the range 1200 to 1900°C in vacuum failed to give pure ThB$_4$ due to the formation of a Th$_x$B$_y$C$_z$ compound at 1250 to 1300°C [30].

Special Preparation Techniques

Thorium borides were obtained by the reaction of thorium oxide and boron oxide, B$_2$O$_3$, in liquid sodium. The reactants were sealed under one atmosphere of argon, the temperature then raised to 870°C within 2 h and maintained for 4 h. The reaction mass was then cooled to 260°C and filtered at 260°C. The filter cake was washed with methanol to remove sodium and any traces of sodium oxide. Lastly the filter cake was broken-up in ordinary atmosphere, added to hot water, and re-filtered [31].

Metal borides were obtained by the reaction of a metal halide and boron to form the metal boride and boron halide. According to this reaction UB$_4$ was prepared from UF$_4$, and this type of reaction was claimed also for the formation of ThB$_4$ [32].

Sintering Behavior and Hot-Pressing

The densification of ThB$_4$ powders by sintering techniques was not successful at temperatures up to 2000°C in vacuum (10^{-5} Torr) due to loss of boron during sintering. In a first stage boron-deficient ThB$_4$ was observed followed by a decrease of 0.46% in the c lattice parameter. In a second stage decomposition was observed with the formation of free metal [26], see also [8, 23].

Hot-pressing of ThB$_4$ powders in tungsten dies at 1250°C using graphite spacers above and below the powder led, after vacuum annealing at 1780°C for 6 h, to a densification of 93 to 95% th.d. (theoretical density 8.45 g/cm^3) [23]. The bulk density and the apparent porosity as results of hot-pressing (2 tsi) in graphite tools [26] are summarized in a figure in [27]. A more dense and purer ThB$_4$ was obtained when a boron nitride separator was used to protect the specimen from oxidation during pressing [26], see also [27].

Single Crystals

Single crystals of ThB$_4$ of a few tenths of a millimeter in size were obtained during synthesis from the elements at 1500°C in vacuum. The crystals formed rectangular parallelepipeds, nearly cubic in habit with very flat faces and sharp edges, and with black metallic luster [20], see also [21].

Enthalpy and Free Energy of Formation

A value of the enthalpy of formation of ThB$_4$ was estimated from consideration of the stability of ThB$_4$ to be $\Delta H_f < -13$ kcal/g-atom boron (¼ ThB$_4$) [33], see also [12, 34, 35]. The free energy of formation was derived from measurements on solid electrochemical cells at 800 to 900°C to be $\Delta G_f^\circ = -52.0$ kcal/g-atom Th [36], see also [7, 12, 22].

13.1.3.2 Crystallographic Properties and Bonding

ThB$_4$ crystals are tetragonal (UB$_4$ type) with Z = 4; the space group is D_{4h}^5-P4/mbm (No. 127) [20, 37], see also [21, 38, 39]. The measured lattice parameters are summarized in Table 59. From this the X-ray density was calculated to the 8.45 g/cm^3 [20, 37], see also [34, 35].

Table 59
Measured Lattice Parameters of ThB$_4$.
Secondary references are given under "Ref." after the semicolon.

a in Å	c in Å	remarks	Ref.
7.256 ± 0.004	4.113 ± 0.002	single crystals	[20, 37]; [5, 8, 12, 18, 27 to 29, 34, 40, 41]
7.258	4.113	hot-pressed	[40]; [12]
7.258 ± 0.002	4.116	powder	[26, 27]; [12]
7.257 ± 0.002	4.097 ± 0.002	arc-melted	
7.258	4.113	B-rich	[8, 42]; [12]
7.257	4.091	Th-rich	
7.265	4.110	hot-pressed	[23, 43]; [12]
7.262	4.096	arc-melted	[44]
7.260	4.120		[45]

It is assumed from measurements on boron-rich and thorium-rich samples that a small homogeneity range in ThB$_4$ cannot be excluded [8]; otherwise it was observed that samples containing 50 to 85 at.% boron showed no difference in the lattice constants of the ThB$_4$

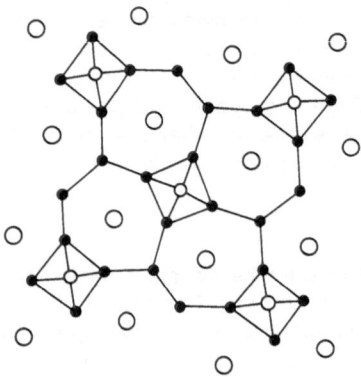

Fig. 40. Structure of ThB₄, schematically [46].

phase [5]. The atomic positions are: 4 Th in: ± (u, ½ + u, 0; ½ − u, u, 0); 4 B(1) in: ± (0, 0, v; ½, ½, v); 4 B(2) in: ± (w, ½ + w, ½; ½ − w, w, ½); 8 B(3) in: ± (x, y, ½; ½ + x, ½ − y, ½; ȳ, x, ½; ½ + y, ½ + x, ½), where u = 0.313 ± 0.002, v = 0.212, w = 0.087, x = 0.170, y = 0.042 [20], see also [37, 41]. Within this structure the boron atoms form a three-dimensional network with characteristic seven-membered rings forming heptagonal prisms around the thorium atoms (see **Fig. 40**) [46], the four thorium atoms lie in the plane with z = constant [20], see also [47].

The atomic distances are: B(1)-B(1) = 1.74 Å, Th-4 B(1) = 2.78 Å, B(1)-4 B(3) = 1.74 Å, Th-4 B(2) = 2.96 Å, B(2)-B(2) = 1.79 Å, Th-2 B(2) = 3.10 Å, B(2)-2 B(3) = 1.79 Å, Th-8 B(3) = 2.84 Å, B(3)-2 B(1) = 1.74 Å, B(3)-B(2) = 1.79 Å, Th-4 Th = 3.74 Å, B(3)-2 B(3) = 1.80 Å, Th-Th = 3.85 Å, Th-2 Th = 4.11 Å [20].

13.1.3.3 Mechanical Properties

Experimental densities of 93 to 95% th.d. (up to 8.0 g/cm³) were obtained after hot-pressing of ThB₄ powders at 1250°C and subsequent annealing at 1780°C in vacuum [23]. A density of 8.4 g/cm³ (porosity less than 1%) after hot-pressing at 1800°C was obtained in [26], see also [27].

Values for the Vicker's hardness (in kg/mm²) are reported to be 2630 (arc-melted specimen) or 2300 (hot-pressed specimen) [26], and 1000 to 1200 [27], also cited in [12]. A value for the Knoop hardness is found to be 2043 (at 100 g load) [24]. The modulus of rupture for hot-pressed specimens is 20000 lb/in² (at 20°C) [26], see also [27].

13.1.3.4 Thermal Properties

The thermal expansion coefficient of ThB₄ was measured from room temperature to 1000°C on hot-pressed specimens and was found to be α = 5.9 × 10⁻⁶°C⁻¹ [26, 27], see also [12]. A value of α = 9.9 × 10⁻⁶°C⁻¹ is given for the temperature range 1100 to 1900°C [12] from [48], where a value in in. per in. · F⁻¹ is given.

ThB₄ melts congruently [8]. The melting temperature was observed to be a little below 2500°C (samples with 5% ThO₂ and an equal amount of ThB₆) [8], also cited in [12]. Melting point above 2500°C [5], about 2500°C (samples prepared by arc-melting) [9], see also [24, 34, 35, 49], or 2210°C, determined in an atmosphere of CO_2 on hot-pressed specimens [26], see also [12, 27].

An empirical equation for the enthalpy of ThB$_4$, measured by drop calorimetry, is $H_1 - H_2 = 0.128\,t - 0.81 \times 10^{-6} \times t^2 + 1008.8 \times t^{-1} - 35.62$ (above 500 F, in Btu/lb, t in °F) [48] or $H - H(0°C) = 0.128\,t - 2.62 \times 10^{-6} \times t^2 + 31.14 \times (t + 18) - 17.51$ (in cal/g, t in °C) [12] from [48]. Values for the specific heat were derived from the slope of the enthalpy versus temperature (see Table 60).

Table 60
Enthalpy and Heat Capacity Data for ThB$_4$ [48].

drop temperature in °F	enthalpy from drop temperature to 32 F in Btu/lb	mean temperature for heat capacity by slope measurement, in °F	heat capacity by slope measurement in Btu · lb^{-1} · F^{-1}
501	30.3	500	0.118
1015	56.4	1000	0.122
1015	96.7	1500	0.124
1512	144.0	2000	0.124
1593	165.4	2500	0.124
2000	217.2	3000	0.124
2520	280.6	3500	0.124
3090	361.6		
3510	399.0		
3555	412.0		

A value of the thermal conductivity at room temperature measured on hot-pressed samples is $K = 0.13\ \text{W} \cdot \text{cm}^{-1} \cdot °\text{C}^{-1}$ (at 25°C) [27], also cited in [12], or $0.16\ \text{W} \cdot \text{cm}^{-1} \cdot °\text{C}^{-1}$ [26]. Measured high-temperature values for the thermal conductivity are summarized in Table 61 (see also **Fig. 41**, p. 124).

Table 61
Thermal Conductivity Data for ThB$_4$.
The specimen was found broken on post-inspection [48].

mean temperature in °F	thermal conductivity in Btu · h^{-1} · ft^{-2} · °F^{-1} · in^{-1}	mean temperature in °F	thermal conductivity in Btu · h^{-1} · ft^{-2} · °F^{-1} · in^{-1}
1184	186.1	2373	240.7
1202	200.1	2375	212.5
1700	178.3	2403	238.1
1700	181.7	2921	247.3
1700	263.6	2924	253.2
1700	244.7	3456	382.5
1700	252.9	3456	387.6
2348	217.2	3477	390.0

References for 13.1 on pp. 135/8

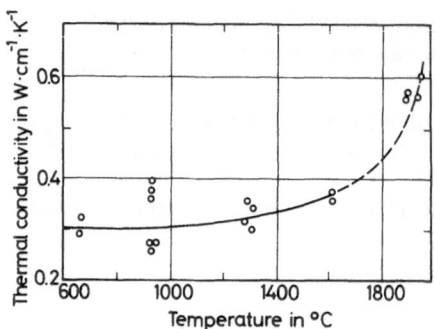

Fig. 41. Thermal conductivity of ThB$_4$ [12] based on values from [48].

Values for the total normal thermionic emittance of ThB$_4$ are given in Table 62 (see also **Fig. 42**).

Table 62
Total Normal Thermionic Emittance of ThB$_4$ [48].

temperature in °F	emittance		temperature in °F	emittance
1571	0.65		2837	0.91
1794	0.66		3172	0.80
1957	0.75		3373	0.76
2192	0.87		3580	0.74
2330	0.94		3739	0.75
2579	0.92			

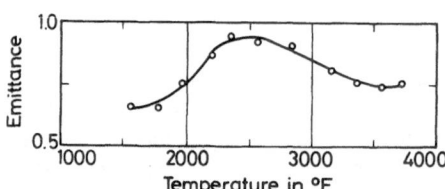

Fig. 42. Total normal emittance of ThB$_4$ [48].

13.1.3.5   Electrical Properties

Electric Resistivity

The specific resistivity of hot-pressed ThB$_4$ specimens at room temperature is $2 \times 10^{-4} \, \Omega \cdot cm$ [27], see also [26], $67 \pm 5 \, \mu\Omega \cdot cm$ (at 300 K, corrected to th.d.) [43], $68 \pm 5 \, \mu\Omega \cdot cm$ (at 300 K, corrected to th.d.) [23]. A linear temperature dependence of the specific resistivity was observed at temperatures ranging from 78 to 700 K, $\varrho_T = 4.5 \times 10^{-2}$ T + 55 (T in K) (see **Fig. 43**) [23], see also [12]. Values for the high-temperature region of 2200 to 3500 F (1200 to 1950°C) are summarized in Table 63 (see also **Fig. 44**) [48], see also [12].

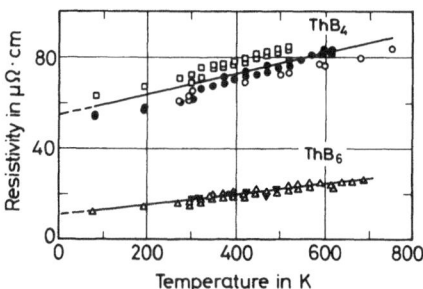

Fig. 43. Electrical resistivity of ThB$_4$ and ThB$_6$ at 78 to 700 K, corrected to theoretical density [23].

Table 63
Electrical Resistivity Data for ThB$_4$ [48].

temperature in °F	resistivity in $\mu\Omega \cdot$ cm	temperature in °F	resistivity in $\mu\Omega \cdot$ cm
2200	2348.6	2855	1856.5
2220	2349.3	3160	1333.7
2230	2352.3	3185	1305.2
2495	2064.3	3185	1284.2
2500	2041.0	3440	1071.1
2505	2063.5	3470	677.3
2830	1822.7	3500	157.5
2840	1862.4	3500	146.3

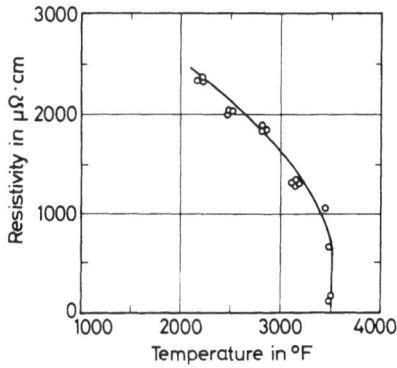

Fig. 44. Electrical resistivity of ThB$_4$ at 2200 to 3500 °F (= 1200 to 1950°C) [48].

References for 13.1 on pp. 135/8

Thermoelectric Power

The temperature coefficient of the thermoelectric power (Seebeck coefficient) of ThB_4 was measured to be S = 2.5 µV/°C (sintered specimen at room temperature) [50], or 5 ± 1 µV/°C (hot-pressed specimen at 300 K) [23, 43], see also [12]. Measured values of thermoelectric voltages within the temperature range of 1765 to 3190°F for specimens in contact with graphite are summarized in [48].

Hall Coefficient

The Hall coefficient of ThB_4 is reported to be R = -1.5×10^{-4} cm^3/Coulomb (at 300 K) [43], or -1.6×10^{-4} cm^3/Coulomb [23], see also [12], corrected to theoretical density. The temperature independence and the negative sign of the Hall coefficient (see **Fig. 45**) indicate that the carriers are electrons in a single band [23], see also [12]. The carrier concentration was calculated from $R = \dfrac{1}{e \cdot n}$ to n = 4.2×10^{22} electrons/cm^3 [23], see also [12], or two electrons per thorium atom [43].

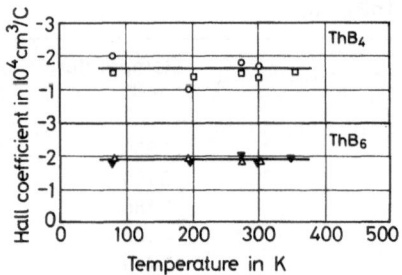

Fig. 45. Hall coefficient versus temperature for ThB_4 and ThB_6, corrected to theoretical density [23].

13.1.3.6   Magnetic Properties

The magnetic susceptibility of the paramagnetic ThB_4 was measured using the Faraday method at 77 to 300 K [23, 44] in magnetic fields of up to 30 kOe [44] and at 80 to 500 K in fields of 2 to 8 Oe [45], see also [51]. Room temperature values of the specific susceptibility are χ_g (in cm^3/g) = 0.04×10^{-6} [23], 3.7×10^{-8} (at 300 K) [44], and 1.204×10^{-6} (at 290 K) [45], see also [51]. The magnetic susceptibility decreases slightly with temperature [23, 45]. The Pauli spin paramagnetic susceptibility, χ_s, was calculated from the equation $\chi_s = (4\, m*\mu_0^2/h^2)\, (3\pi^2 n)^{1/3}[1-(m^2/3m*^2)]$ where μ_o = Bohr magneton, m = electron mass, m* = effective electron mass (two times the electron mass), h = Planck's constant, n = electron concentration. This calculated value is $\chi_s = 0.7 \times 10^{-6}$ cm^3/g, whereas the measured (total) susceptibility was 0.04×10^{-6} cm^3/g, which indicates appreciable contributions from other sources [23]. A Fermi level of 2.2 ± 0.5 eV was calculated using a free-electron theory with the assumption that the mean free time between scattering events is constant [23].

13.1.3.7   Chemical Reactions

For chemical behavior of ThB$_4$ reported in the older literature, see "Thorium" 1955, pp. 295/6.

On Heating

ThB$_4$, gray in color, decomposes on annealing at 2000°C under vacuum conditions in two steps. Loss of boron leads in the first step to a boron-deficient substance combined with a decrease of the lattice parameter c. Thorium metal is formed in the second step [26, 27], also referred to in [12].

With Elements

Powdered ThB$_4$ is stable in dry hydrogen up to 600°C, hot-pressed material up to 650°C [26, 27], also referred to in [12]. The oxidation of ThB$_4$ in oxygen is reported to be negligible up to 600°C; a hard glaze of B$_2$O$_3$ is formed at higher temperatures which presumably forms a protective layer on the surface of the specimen [26, 27], also referred in [12]. A linear oxidation rate of 0.22 mg · cm^{-2} · h^{-1} was observed at 1100°F in dry air [24], see also [52]. There was no formation of nitrides observed when ThB$_4$ (powdered or hot-pressed samples) was heated in nitrogen up to 1000°C [26, 27], also referred to in [12]. ThB$_4$ is considered from preparation techniques to be unstable in the presence of carbon [33], see also [12], but there was no apparent reaction observed between carbon and ThB$_4$ (powdered or hot-pressed samples) below 2000°C [26, 27], also referred to in [12].

With Compounds

ThB$_4$ is oxidized in an atmosphere of CO$_2$ at 680°C (powdered samples). The oxidation is more rapid at 600°C than at 1000°C (hot-pressed samples), which indicates the formation of a protective oxide layer on the surface of the samples [26, 27], also referred in [12]. An oxidation rate of 0.08 mg · cm^{-2} · h^{-1} was observed at 1100°F [24], see also [52]. ThB$_4$ is reported to be insoluble in water [12]. A small corrosion rate of 0.6 × 10^{-2} mg · cm^{-2} · h^{-1} was observed at 90°C [9], see also [24, 52]. ThB$_4$ is reported to be soluble in aqueous HNO$_3$, HCl, and hot H$_2$SO$_4$. It decomposes in acidic solutions of potassium iodate and Ce^{III} sulfate [12]. A small corrosion rate of 1.6 × 10^{-2} mg/cm^2 per day was observed with ThB$_4$ in contact to NaK at 650°C [9], see also [24, 52].

ThB$_4$ and CeB$_4$ form a complete series of solid solutions [8], see also [42]. Solid solutions apparently are formed with ThO$_2$ [5]. A complete series of solid solutions is formed with UB$_4$ [40], see also [8]. A eutectic composition between β-Th and ThB$_4$ with a eutectic temperature of 1450°C [8], see also [42], or 1550°C [5] is placed close to pure thorium, see also [38, 53]. A further eutectic composition is formed between ThB$_4$ and ThB$_6$ with a melting point of about 2400°C [8, 9].

13.1.4   Thorium Hexaboride, ThB$_6$

13.1.4.1   Formation and Preparation

ThB$_6$ was already prepared in 1905 from melts with hyperstoichiometric amounts of boron [17]. Detailed information of the older literature is given in "Thorium" 1955, pp. 295/6.

                References for 13.1 on pp. 135/8

Synthesis from the Elements

The reaction of thorium powder and boron powder in stoichiometric amounts starts at 850°C [6] or at 950°C [5] under vacuum conditions when a sharp rise in temperature is observed [5] with the formation of ThB_4. The excess boron reacts at about 1000°C with ThB_4 forming ThB_6 [6]. The reaction was carried out in high vacuum furnaces (10^{-4} to 10^{-5} Torr) [8] or in a high-frequency induction furnace (10^{-6} Torr) [6] with the reactants pressed into pellets and placed in a tungsten crucible [8]. The reaction was completed at 1400°C within 3 h (with 10% excess of boron) [23], see also [22], at 1300 to 1450°C within 2 h [6], see also [12, 54, 55], or at 1600°C within 10 h [8]. For reaction at temperatures of up to 2400 K, see [56, 57]. Boron-rich samples of ThB_6 were prepared with atomic ratios $6 \leqslant B/Th \leqslant 24$ under high-vacuum conditions at 1300°C. The limit of the homogeneity range of the $Th_{1-x}\square_xB_6$ compounds was found at $x = 0.22$ from X-ray diffraction studies [10].

Reduction of ThO_2 with Boron (and Carbon and/or B_4C)

Mixed powders of thorium dioxide, boron, and carbon, pressed into pellets (5 tsi) and placed in a graphite crucible, were heated in an induction furnace in an atmosphere of hydrogen at 1500°C for 2 h or at 1800°C for 1 h to obtain ThB_6. There were strong indications, especially for the reaction at 1800°C, that boron rather than carbon acted as the reducing agent with liberation of B_2O_3 [28], see also [29]. ThB_6 was also readily formed at 1500°C and 1800°C when the initial powder mixture contained no carbon. In this case the reaction product contained 1 to 2% carbon picked up from the graphite crucible. ThB_6 was equally well obtained if ZrB_2 crucibles were used [28], see also [29, 35, 58], also referred in [12]. The thermal reduction of ThO_2 with boron carbide, B_4C, and carbon (lamp-black) under vacuum conditions (10^{-3} Torr), with the powders mixed and pressed into pellets, was observed to be very slow at 1300 to 1400°C ($2 ThO_2 + 3 B_4C + C = 2 ThB_6 + 4 CO$). Complete reaction was obtained at 1800 to 1900°C within 35 to 45 min, or at 1600°C within 1 h. The ThB_6 formed contained 0.09 to 0.18% free carbon [30, 59], also referred to in [12].

Special Preparation Techniques

ThB_6 was prepared crucible-free by a special floating-zone technique from ThO_2 and boron ($ThO_2:B = 1:8$). ThO_2 and boron were thoroughly mixed (wetted with benzene), pressed into rods of 25 mm length, and placed into the vacuum chamber of the electron-beam melting furnace. The reaction to give ThB_6 is reported to be fast and the method remarkably simple [60].

ThB_6 was successfully prepared by electrolytic reduction of ThO_2 dissolved in a melt of boron oxide (B_2O_3) with alkali and alkaline earth metal oxides and fluorides. The electrolysis was carried out at 900 to 1100°C [61, 62], see also [63], also referred to in [12].

Thorium borides were also obtained by the reaction of ThO_2 with B_2O_3 in liquid sodium. The reactants were sealed under argon, the temperature then raised to 870°C within 2 h, and maintained for 4 h. The reaction mass was then cooled to 260°C and filtered at 260°C. The filter cake was washed with methanol to remove sodium and any traces of sodium oxide. Lastly the filter cake was broken up in ordinary atmosphere, added to hot water, and refiltered [31].

Sintering Behavior and Hot-Pressing

The densification of ThB$_6$ powders by sintering techniques is difficult due to the high vapor pressure at sintering temperatures, which prevents a reduction of the porosity. The porosity remains up to 30% after sintering in vacuum (1.333 N/m^2). The porosity decreases to 8% after sintering in an atmosphere of hydrogen at normal pressure [59].

ThB$_6$ powders were successfully hot-pressed at a pressure of 20.7 MN/m^2 at 2000°C within 5 min. The remaining porosity was measured to be 0.7%, and the density of the specimens was 7.05 g/cm^3 (theoretical density 7.10 g/cm^3) [59], also cited in [12]. Powders which were hot-pressed at 1300°C, then annealed at 1725 to 1750°C in vacuum and pelleted formed ThB$_4$ on the surface due to decomposition of the ThB$_6$. The density was found to be only 75% th. d. [23].

Single Crystals

Boron-rich single crystals of 0.5 to 2 mm in size were grown at 1550°C within 24 h by an Al-Ga flux growth method in a special high-temperature thermal reactor (see **Fig. 46**) [64].

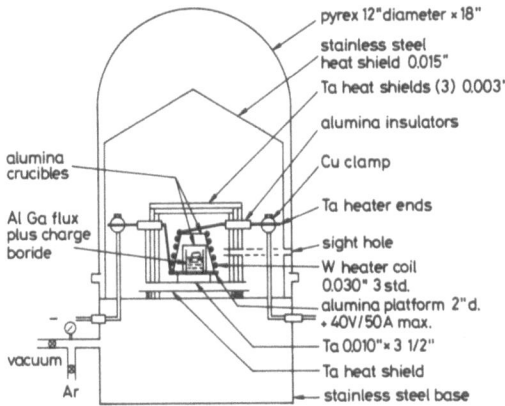

Fig. 46. High-temperature thermal reactor for the preparation of ThB$_6$ single crystals using an Al-Ga flux method [64].

Enthalpy and Free Energy of Formation

The enthalpy of formation of ThB$_6$ has been estimated to be $\Delta H_f^\circ < -11$ kcal/g-atom boron ($^1/_6$ ThB$_6$) [33], see also [34, 35] or $\Delta H_f^\circ = -60$ kcal/mol [65], from which value the heat of atomization was calculated to be 142 kcal/g-atom [65]. The free energy of formation was measured using high-temperature solid emf cells at 800 to 950°C and found to be $\Delta G_f^\circ = -55$ kcal/g-atom thorium [22] or -54.4 kcal/mol [36], see also [66], also cited in [12]. An estimated value of the free energy of formation is reported to be less than -54 kcal/mol [48], also referred to in [12].

13.1.4.2   Crystallographic Properties and Bonding

ThB_6 crystallizes in the cubic system (CaB_6 type) with $Z = 1$; the space group is $Pm3m-O_h^1$ (No. 221) [10], also referred to in [41], see also [34, 35, 38, 67, 68, 69, 70]. The measured lattice parameters are summarized in Table 64. From this the X-ray density is calculated to be 7.12 g/cm³ [71] or 7.10 g/cm³ [6, 55].

Table 64
Measured Lattice Parameters of ThB_6.
References after the semicolon are secondary ones.

a in Å	remarks	Ref.
4.15		[70]; [18]
4.16		[72]; [28]
4.113		[73, 74]; [8, 12, 28, 40]
4.110	sintered	[28, 29, 55]; [12]
4.101	hot-pressed	[59, 71]; [12]
4.111	hot-pressed	[40]; [12]
4.111 ± 0.002	hot-pressed	[23, 26]; [12]
4.110	B-rich	[8, 42]; [12]
4.108	Th-rich	[8, 42]; [12]
4.108	sintered	[23, 43]; [12]
4.106		[65, 75]; [12]
4.1105 ± 0.0005	B-rich	[6, 10]; [12]
4.1125 ± 0.0005	Th-rich	[10]; [12]
4.1132		[41]

The temperature dependence of the lattice parameter was measured between room temperature and 1628 °C with a high-temperature X-ray camera [65]. The following values are reported from these measurements [12]:

t in °C	room temperature	1010	1212	1420	1628
a in Å.	4.104	4.139	4.149	4.158	4.168

The temperature dependence of the lattice parameter a is also given as $a = 4.1095\,(1 + 7.37 \times 10^{-6}\,t + 1.66 \times 10^{-9}\,t^2)$ [80].

It was assumed from the existence of boron-rich and thorium-rich samples that ThB_6 exists within a homogeneity range [8]. A special investigation performed on samples with B/Th ratios in the range of $6 \leqslant B/Th \leqslant 24$ led to lattice parameters ranging from $a = 4.1105 \pm 0.0005$ to $a = 4.1125 \pm 0.0005$ Å for samples of $Th_{1-x}\square_xB_6$ with $x = 0$ and $x = 0.22$, respectively [10]. The atomic positions are: origin at center (m3m); 1 Th in 0,0,0; 6 B in x,½,½; ½,x,½; ½,½,x; x̄,½,½; ½,x̄,½; ½,½,x̄ [41]. Within this structure, the six boron atoms form a regular octahedron which is centered about the body center of the cell and with its axes parallel to the cell axes. The thorium atoms are located in the corner positions of the cells (see **Fig. 47**) [41, 55], see also [76 to 78]. The shortest atomic distances are Th-Th = 4.16 Å [72], 4.110 Å [79]; Th-B = 3.07 Å [72], 3.028 Å [79]; B-B = 1.72 Å [72], 1.703 Å [79].

Characteristic temperatures of ThB_6, determined from the reduction of the peak intensity in the range from room temperature to 1000 °C, were found to be $\Theta_{Th} = 245 \pm 12$ K using a

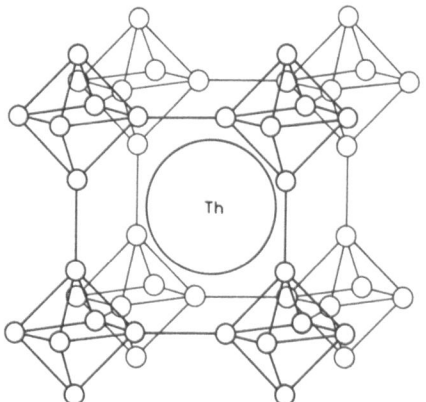

Fig. 47. Structure of ThB$_6$, schematic [55].

heavy mass model and $\Theta = 398$ K using an average mass model [75]. From this data the mean square vibrational amplitude at room temperature was calculated to be $\mu^2 = 0.016$ Å^2 [75]. Further values are reported to be $\Theta = 600$ K [71] or 576 K [81], and $\sqrt{\bar{\mu}^2} = 0.045$ Å (for boride), $\sqrt{\bar{\mu}^2} = 0.111$ Å (for thorium) [71], also cited in [59, 82].

13.1.4.3 Mechanical Properties

Experimental densities of 7.05 g/cm^3 (theoretical density 7.10 g/cm^3) were obtained after hot-pressing of ThB$_6$ powders at a pressure of 20.7 MN/m^2 at 2000°C within 5 min [59]. Experimental densities of 75% th.d. were obtained after annealing hot-pressed powders at 1725 to 1750°C [23].

A value of the microhardness of ThB$_6$ is reported to be 1740 ± 123 kg/mm^2 (at 20 g load) measured at hot-pressed specimens [30], also cited in [12, 34]. The Vicker's hardness was measured at arc-melted samples to 2590 kg/mm^2 [26], also cited in [12].

13.1.4.4 Thermal Properties

The thermal expansion of ThB$_6$ was measured by X-ray diffraction methods. The following values were obtained for the thermal expansion coefficient, α (average values): $\alpha = 7.8 \times 10^{-6}$°C^{-1} at 20 to 800°C [71], see also [59, 83], also referred to in [12, 65], $\alpha = 11.3 \times 10^{-6}$°C^{-1} at 900 to 1700°C (see **Fig. 48**, p. 132) [65], see also [84]. The experimental curve of the thermal coefficient versus temperature was fitted to the equation $\alpha = \alpha_0 + \alpha_1 t^2 + \ldots$ [85], see also [80]. Coefficients α and constants are given in Table 65, p. 132.

The following melting points of ThB$_6$ were measured: 2150°C [30], see also [59, 71, 86], also referred to in [12], 2190°C [9], see also [24], 2195°C [11], see also [49], also referred to in [12, 34, 35, 53, 65]. A melting point of 2180°C was calculated from X-ray determination of the characteristic temperature, Θ, and the vibrational amplitude, μ [71]. An investigation indicating congruent melting of ThB$_6$ at 2450°C and showing a ThB$_6$-boron eutectic at 2050°C [8] possibly reflects the presence of free boron in the ThB$_6$ samples at the other determinations [12].

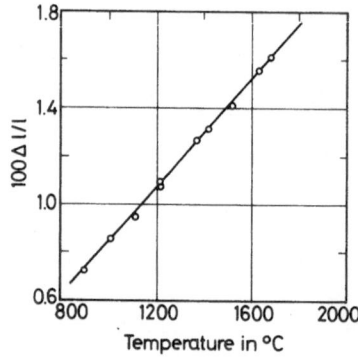

Fig. 48. Thermal expansion versus temperature for ThB_6 [65].

Table 65
Measured Thermal Expansion Coefficients of ThB_6 and the Constants α_0 and α_1 of the Equation $\alpha = \alpha_0 + \alpha_1 t^2$ in Terms of α_0/α and α_1/α [85].

temperature in K	$\alpha \times 10^6$ in °C^{-1}	$100\ \alpha_0/\alpha$	$100\ \alpha_1/\alpha$	temperature in K	$\alpha \times 10^6$ in °C^{-1}	$100\ \alpha_0/\alpha$	$100\ \alpha_1/\alpha$
273	7.37	100	0	523	8.18	89.88	10.12
293	7.44	99.11	0.89	573	8.34	88.09	11.91
323	7.53	97.80	2.20	673	8.67	84.73	15.27
373	7.70	95.69	4.31	773	8.99	81.62	18.38
423	7.86	93.67	6.33	873	9.31	78.72	21.28
473	8.02	91.74	8.26	973	9.63	76.03	23.97

The heat capacity of ThB_6 was measured within the low-temperature region at 2 to 12 K by adiabatic calorimetry (see **Fig. 49**) [54], see also [55]. The results are expressed by the equation $C = \gamma \cdot T + \beta \cdot T^3$ with the constants $\gamma = 4.8$ mJ $\cdot$ mol$^{-1} \cdot$ K^{-2} (electronic contribution) and $\beta = 0.3$ mJ $\cdot$ mol$^{-1} \cdot$ K^{-1}, from which a Debye temperature of $\Theta_D = 188$ K was derived [54], see also [12, 55]. Two different values were derived for the entropy of ThB_6 from X-ray determination of characteristic temperature data: $S = 42.6$ cal $\cdot$ (g-atom Th)$^{-1} \cdot$ K^{-1} (using an average mass model) or $S = 27.8$ cal $\cdot$ (g-atom Th)$^{-1} \cdot$ K^{-1} (using a heavy mass model) [75], see [87].

A value of 44.9 W $\cdot$ m$^{-1} \cdot$ K^{-1} is reported for the thermal conductivity of ThB_6 [59], see also [12, 82].

The current density, I, of the thermionic emission of ThB_6 is given as log $I/T^2 = $ log A $-$ 11 600 $\varphi/2.303$ T with $\varphi = 2.92$ eV (work function) and A $= 0.53$ A $\cdot$ cm$^{-2} \cdot$ K^{-2} [88], see also [11], also cited in [12]. Further values are reported: effective work function $= 2.92$ eV (at 1700 K) [59] or 2.86 eV [30], also cited in [12, 34]; coefficient of radiation $= 0.69$ to 0.70 [30], also cited in [12, 34]; radiation factor $= 0.74$ [59], see also [12].

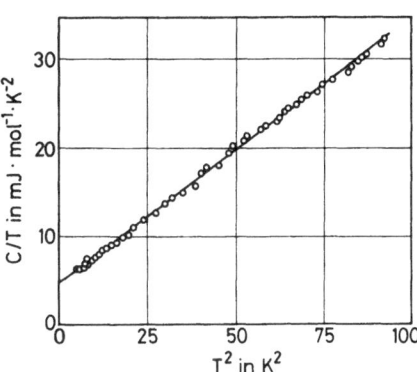

Fig. 49. Low-temperature heat capacity of ThB$_6$, expressed as C/T $= \gamma + \beta T^2$ [55].

13.1.4.5 Electrical Properties

Values for the room temperature specific electric resistivity of ThB$_6$ specimens are reported to be $37 \times 10^{-6}\,\Omega \cdot$ cm [30], also cited in [34], 14.8 $\mu\Omega \cdot$ cm [89], see also [59, 79, 90], also referred to in [12, 65], $34 \pm 1\,\mu\Omega \cdot$ cm (at 300 K), corrected to theoretical density [43], or 18 $\mu\Omega \cdot$ cm (at 300 K), corrected to theoretical density [23]. A linear temperature dependence of the specific resistivity was observed at temperatures ranging from 78 to 700 K with $\varrho = 11 + 2.25 \times 10^{-2}\,$T (see Fig. 43, p. 125) [23], also referred to in [12], or $d\varrho/dT = 2.31 \times 10^{-3}\,°C^{-1}$ at 0 to 100°C [89], see also [59, 79, 90].

Superconductivity was observed at ThB$_6$ with a transition temperature T$_c$ of less than 1.28 K [74], see also [58].

The temperature coefficient of the thermoelectric power (Seebeck coefficient) was measured to be S (in μV/°C) $= 2.5$ (at 20 to 80°C, in contact with copper) [30], also cited in [12, 34], 4.38 (in contact with copper) [50], -0.6 [89], see also [59, 79, 90], also referred to in [12], -5 ± 1 (at 300 K) [23, 43, 55], also referred to in [12].

The Hall coefficient of ThB$_6$ is reported to be R (in cm^3/Coulomb) $= -2.18 \times 10^{-4}$ (at room temperature) [89], see also [59, 79, 90], also referred to in [12], -2.0×10^{-4} (at 300 K, corrected to theoretical density) [43], -1.9×10^{-4} (at 78 to 350 K) [23], also referred to in [12]. The temperature independence and the negative sign of the Hall coefficient (see Fig. 45, p. 126) indicate that the carriers are electrons in a single band [23]. The carrier concentration was calculated from R $= \dfrac{1}{e \cdot n}$ to be 3.1×10^{22} electrons/cm^3 [23], or 1.99 electrons/Th atom (effective current carrier) [89], see also [59, 79, 90], and a current carrier mobility of 14.7 cm$^3 \cdot$ V$^{-1} \cdot$ s^{-1} was found [89], see also [59, 79, 90].

X-ray valence band spectra and photoelectron spectra are reported for several rare earth hexaborides and ThB$_6$. The results were discussed in terms of the valence and conduction bands and of the life-time broadening of the 4f lines [91].

13.1.4.6 Magnetic Properties

ThB$_6$ shows a temperature-independent, slightly diamagnetic behavior [23, 55]. The specific magnetic susceptibility is $\chi_g = 13 \times 10^{-6}$ cm^3/g [59], also referred to in [12], -0.08×10^{-6} cm^3/g

References for 13.1 on pp. 135/8

(at 300 K) [23], also referred to in [12]. An effective magnetic moment of $\mu_{eff} = 0.18\,\mu_B$ is reported for ThB_6 in [59].

The Pauli spin paramagnetic susceptibility, χ_s, was calculated from the equation $\chi_s = (4m^* \cdot \mu_0^2/h^2)(3\pi^2 n)^{1/3}[1-(m^2/3m^{*2})]$, where μ_0 = Bohr magneton, m = electron mass, m* = effective electron mass (1.6 times the electron mass), h = Planck's constant, n = electron concentration. The calculated value is $\chi_s = 0.14 \times 10^{-6}\,cm^3/g$, whereas the measured (total) susceptibility was $-0.08 \times 10^{-6}\,cm^3/g$, which indicates appreciable contributions from other sources [23]. A Fermi level of 2.2 ± 0.5 eV was calculated using a free-electron theory with the assumption that the mean free time between scattering events is constant [23].

13.1.4.7 Chemical Reactions

For chemical reactions of ThB_6 reported in the older literature, see "Thorium" 1955, p. 296.

ThB_6, red to red-violet in color, decomposes to ThB_4 at temperatures above 1650°C [6]. Otherwise, congruent melting of ThB_6 was observed at 2450°C [8]. ThB_6 was found to be stable in the presence of oxygen [11], also referred to in [12], and in the presence of graphite [33]. No reaction of ThB_6 was observed with moisture [11], also cited in [12]. For reactions with acids see Table 66. ThB_6 is stable against solutions of alkali hydroxides [61], see also [30]. Vigorous reaction is reported with PbO_2 and Na_2O_2 [61], see also [30]. ThB_6 is easily decomposed by fusing with carbonates and hydrogen sulfates of alkali metals [61], see also [30]. A ThB_6-boron eutectic is reported with a melting point of about 2050°C [8], see also [12].

Table 66
Reactions of ThB_6 with Acids.
References after the semicolon are secondary ones.

acid	reaction	Ref.
HCl (1:1)	82% insoluble, residue after 2 h of dissolution	[92]
concentrated HCl	no reaction	[11, 61]; [12, 30]
HF	no reaction	[61]; [30]
dilute H_2SO_4	no reaction	[61]; [30]
concentrated H_2SO_4	easily decomposed on heating	[61]; [30]
H_2SO_4 (1:1) + 5 drops of HNO_3	complete dissolution on heating	[92]
HNO_3 (1:1)	complete dissolution on heating	[92]
concentrated HNO_3	easily decomposed on heating	[61]; [30]
aqua regia	complete dissolution in the cold or on heating	[92]

13.1.5 Thorium Dodecaboride, ThB_{12}

ThB_{12} was prepared by hot-pressing stoichiometric amounts of thorium and boron at a pressure of 65 kbar at 1660°C with the reactants placed in boron nitride crucibles. In every case ThB_6 was present concentrated at the end of the high-temperature reaction chamber. Pure ThB_{12}, silvery in color, was found in the center of the samples [13, 93].

ThB$_{12}$ crystallizes in the cubic system (UB$_{12}$ type). A lattice parameter of a = 7.612 ± 0.001 Å was determined from X-ray diffraction using a Debye-Scherrer camera [13, 93].

The clean separation of ThB$_6$ and ThB$_{12}$ at the above hot-pressing conditions indicates that ThB$_{12}$ may melt incongruently to ThB$_6$ and a melt at high pressures and at a fairly low temperature [13].

13.1.6 Thorium Octadecaboride, ThB$_{18}$

Thorium borides with a B/Th ratio >6 were prepared from thorium and boron powders, pressed into pellets and heated in a high-frequency induction furnace in vacuum. ThB$_4$ and then ThB$_6$ is formed at increasing temperature as observed for ThB$_6$. ThB$_6$ then reacts with the excess boron at temperatures above 1300°C when a new phase, ThB$_{18}$, is formed, brown in color. The observed density of ThB$_{18}$ is 4.42 g/cm^3 [6]. The structure of the compound ThB$_{18}$ has not yet been determined. The d-values and relative intensities, observed from an X-ray diffraction study, are given in the original paper [6].

The compound ThB$_{18}$ disproportionates at 1600°C forming ThB$_6$ and a thorium "hecto-boride" phase, δ (B/Th = 70). This disproportionation was observed to be reversible; at 1500°C the phase ThB$_{18}$ was re-formed: γ (ThB$_{18}$) ⇌ ThB$_6$ + δ [6].

13.1.7 Thorium "Hectoboride", ThB$_{66}$

Thorium "hectoborides" were prepared by argon-arc melting of thorium and boron [14, 15] and were also observed from thorium and boron powders when pressed into pellets and heated up to above 1600°C in a high-frequency induction furnace in vacuum [6]. Melted specimens with a B/Th ratio of 12:1 and 100:1 were observed to consist of two phases forming Th + 12 B → ThB$_6$ + hectoboride and Th + 100 B → hectoboride + boron [14]. A single crystal of ThB$_{66}$ was obtained on slowly cooling a liquid corresponding to a B/Th ratio of 70:1 [94]. ThB$_{66}$ crystallizes in the cubic system (YB$_{66}$ type), Z = 24; the space group is Fm3c-O$_h^6$ (No. 226) [6, 94]. The measured lattice parameters are a = 23.46 ± 0.01 Å (sintered two-phase sample) [6], 23.53 ± 0.01 Å (single crystal) [94], 23.518 ± 0.003 Å (arc-melted sample) [14]. The X-ray density was calculated to be 2.92 g/cm^3 compared to an observed density of 2.98 g/cm^3 [6]. Within this structure the unit cell contains 1584 boron atoms, 1248 of which are placed in 13 B$_{12}$(B$_{12}$)$_{12}$ icosahedrons. The remaining boron atoms are distributed statistically in channels which are formed within the three-dimensional network of the B$_{12}$(B$_{12}$)$_{12}$ units. These boron atoms form non-icosahedral cages. The thorium atoms are inserted in the channels [94], see also [6]. The boron-to-boron distances within the icosahedra range from 1.749 to 1.911 Å. The distances between the icosahedra vary from 1.598 to 1.837 Å, the thorium to boron distances from 2.71 to 2.84 Å [94]. The atomic positions and the bond distances are given in detail in [94].

The microhardness (Knoop) of ThB$_{66}$ was measured at arc-melted specimens to be 2310 ± 50 kp/mm^2 at 100 g load [14].

References for 13.1:

[1] L. H. Andersson, R. Kiessling (Acta Chem. Scand. **4** [1950] 160/4). − [2] W. T. Ziegler, R. A. Young (Phys. Rev. [2] **90** [1953] 115/9). − [3] H. Blumenthal (Powder Met. Bull. **7** [1956] 79/81). − [4] G. V. Samsonov, V. S. Neshpor (Redk. Metal. Splavy Tr. 1st Vses. Soveshch.,

Moscow 1957 [1960], pp. 392/417; C.A. **1961** 4314). – [5] L. Brewer, D. L. Sawyer, D. H. Templeton, C. H. Dauben (J. Am. Ceram. Soc. **34** [1951] 173/9).

[6] J. Etourneau, R. Naslain (Compt. Rend. C **266** [1968] 1452/5). – [7] M. H. Rand (At. Energy Rev. Spec. Issue No. 5 [1975]). – [8] P. Stecher, F. Benesovsky, H. Nowotny (Planseeber. Pulvermet. **13** [1965] 37/46). – [9] N. M. Griesenauer, M. S. Farkas, F. A. Rough (BMI-1680 [1964] 1/32; C.A. **62** [1965] 2454). – [10] J. Etourneau, R. Naslain, S. La Placa (J. Less-Common Metals **24** [1971] 183/93).

[11] J. M. Lafferty (J. Appl. Phys. **22** [1951] 299/309). – [12] S. Peterson (ORNL-4503-Vol. 4 [1971] 1/13; C.A. **78** [1973] No. 139584). – [13] J. F. Cannon, P. B. Farnsworth (J. Less-Common Metals **92** [1983] 359/68). – [14] K. Schwetz, P. Ettmayer, R. Kieffer, A. Lipp (J. Less-Common Metals **26** [1972] 99/104). – [15] K. Schwetz, P. Ettmayer, R. Kieffer, A. Lipp (Radex Rundschau **1972** 257/65; C.A. **78** [1973] No. 8801).

[16] K. A. Gingerich (High Temp. Sci. **1** [1969] 258/67). – [17] Binet du Jassonneix (Compt. Rend. **141** [1905] 191/3). – [18] L. I. Katzin (Natl. Nucl. Energy Ser. Div. IV B **14** [1954] 66/102). – [19] Binet du Jassonneix (Ann. Chim. Phys. [8] **17** [1909] 145/217; C.A. **1909** 2096). – [20] A. Zalkin, D. H. Templeton (Acta Cryst. **6** [1953] 269/72).

[21] A. Zalkin, D. H. Templeton (AECD-2762 [1949] 1/4; N.S.A. **4** [1950] No. 1415). – [22] Brookhaven National Laboratories (BNL-929 [1965] 83/105, 89; N.S.A. **20** [1966] No. 20177). – [23] A. B. Auskern, S. Aronson (J. Chem. Phys. **49** [1968] 172/6). – [24] M. S. Farkas, A. A. Bauer, R. F. Dickerson (BMI-1568 [1962] 1/20; C.A. **56** [1962] 13732). – [25] Anonymous (Reactor Fuel Process. **8** No. 1 [1964/65] 50/64; N.S.A. **19** [1965] No. 23069).

[26] K. J. Matterson, H. Jones (Trans. Brit. Ceram. Soc. **60** [1961] 475/93). – [27] K. J. Matterson, H. J. Jones, N. C. Moore (Plansee Proc. 4th Semin., Reutte/Tyrol, Austria, 1961 [1962], pp. 329/63; N.S.A. **17** [1963] No. 1847). – [28] B. Post, D. Moskowitz, F. W. Glaser (J. Am. Chem. Soc. **78** [1956] 1800/2). – [29] B. Post, D. Moskowitz, F. W. Glaser (Plansee Proc. 2nd Semin., Reutte/Tyrol, Austria, 1955 [1956], pp. 173/86; C.A. **1956** 13637). – [30] G. V. Samsonov, O. N. Zorina (Zh. Neorgan. Khim. **1** [1956] 2260/3; Russ. J. Inorg. Chem. **1** No. 10 [1956] 90/3; C.A. **1957** 8565).

[31] Mine Safety Applicances Co. (Brit. 955730 [1964]; C.A. **61** [1964] 1537). – [32] Reactor Centrum Nederland (Brit. 1257544 [1971]; N.S.A. **26** [1972] No. 15609). – [33] L. Brewer, H. Haraldsen (J. Electrochem. Soc. **102** [1955] 399/406). – [34] G. V. Samsonov, K. I. Portnoi (Splavy na Osnove Tugoplavkikh Soedinenii [Alloys Based on Refractory Compounds], Gos. Izd. Obor. Prom., Moscow 1961, pp. 1/387; FTD-TT-62-430 [1962]; N.S.A. **16** [1962] No. 27640). – [35] G. V. Samsonov, Yu. B. Paderno (Boridy Redkozemel'nykh Metallov [Borides of Rare Earth Metals], Izd. AN Ukr. SSR, Kiev 1961, pp. 1/104; AEC-TR-5264 [1963] 1/104; N.S.A. **17** [1963] No. 2959).

[36] S. Aronson, A. Auskern (BNL-9167 [1964] 1/10; C.A. **63** [1965] 17222). – [37] A. Zalkin, D. H. Templeton (J. Chem. Phys. **18** [1950] 391). – [38] L. Brewer, D. L. Sawyer, D. H. Templeton, C. H. Dauben (AECD-2823 [1959] 1/9; N.S.A. **4** [1950] No. 3419). – [39] A. Zalkin, D. H. Templeton (AECD-3080 [1950] 1/22; UCRL-522 [1950] 1/22; N.S.A. **5** [1951] No. 3083). – [40] L. Toth, H. Nowotny. F. Benesovsky, E. Rudy (Monatsh. Chem. **92** [1961] 945/8; C.A. **56** [1962] 9478).

[41] D. J. Lam, J. B. Darby Jr., M. V. Nevitt (Actinides Electron. Struct. Relat. Prop. **2** [1974] 119/84). – [42] F. Benesovsky, P. Stecher, H. Nowotny, W. Rieger (Colloq. Intern. Centre Natl. Rech. Sci. [Paris] No. 157 [1967] 419/30). – [43] A. Auskern, S. Aronson (BNL-50082 [1967] 126/8; N.S.A. **22** [1968] No. 38510). – [44] K. H. J. Buschow, J. H. N. Creyghton (J. Chem. Phys. **57** [1972] 3910/4). – [45] V. I. Chechernikov, N. K. Ali-Zade, M. G. Ramazanzade, V. K. Slovyanskikh, L. M. Lisitsyn, N. G. Guseinov (Fiz. Tverd. Tela [Leningrad] **16** [1974] 1056/9; Soviet Phys. Solid State **16** [1974] 681/2; C.A. **81** [1974] No. 7751).

[46] S. Andersson, L. Stenberg (Z. Krist. **158** [1982] 133/9). – [47] Yu. B. Paderno (Bor Poluch. Strukt. Svoistva Mater. 4th Mezhdunar. Simp. Boru, Tbilisi, USSR, 1972 [1974], pp. 97/

105; N.S.A. **33** [1976] No. 1159). — [48] Southern Research Institute, Birmingham, Ala. (ASD-TDR-62-765 [1963] 1/420; AD-298061 [1963] 1/420; N.S.A. **17** [1963] No. 16636). — [49] T. A. Badaeva (Str. Splavov Nek. Sist. Uranom Toriem **1961** 339/57 [The Structure of Alloys of Certain Systems Containing Uranium and Thorium]; AEC-TR-5834 [1963] 321/36; N.S.A. **16** [1962] No. 30877). — [50] G. V. Samsonov, N. S. Strel'nikova (Ukr. Fiz. Zh. **3** [1958] 135/8; C.A. **1958** 19309).

[51] V. I. Chechernikov, V. A. Pletyushkin, A. V. Pechennikov, V. I. Nedel'ko, N. Kh. Ali-Zade, Z. B. Chachkhiani, V. K. Slovyanskikh, R. N. Kuz'min (Tr. Mezhdunar. Konf. Magn., Moscow 1973 [1974], Vol. 6, pp. 49/53; C.A. **86** [1977] No. 132427). — [52] M. S. Farkas, A. A. Bauer, R. F. Dickerson (Trans. Am. Nucl. Soc. **5** [1962] 244). — [53] O. von Goldbeck (At. Energy Rev. Spec. Issue No. 5 [1975]). — [54] J. P. Mercurio, J. Etourneau, R. Naslain (Compt. Rend. B **268** [1969] 1766/9). — [55] J. Etourneau, J. P. Mercurio, R. Naslain, P. Hagenmuller (J. Solid State Chem. **2** [1970] 332/42).

[56] Temple University Research Institute, Philadelphia, Penna. (TID-18684 [1963] 1/23; N.S.A. **17** [1963] No. 25164). — [57] A. D. Kirshenbaum, A. V. Grosse (TID-18951 [1963] 1/40; N.S.A. **17** [1963] No. 28828). — [58] K. Hiebl, M. J. Sienko (Inorg. Chem. **19** [1980] 2179/80). — [59] G. V. Samsonov, Yu. B. Paderno, V. S. Fomenko (Poroshkovaya Met. **1963** No. 6, pp. 24/31; Soviet Powder Met. Metal Ceram. **1963** No. 6, pp. 449/54; C.A. **60** [1964] 10192). — [60] Z. Ban, M. S. Sikirica (New Nucl. Mater. Incl. Non-Metal. Fuels. Proc. Conf., Prague 1963, Vol. 2, pp. 175/82; C.A. **60** [1964] 10192).

[61] L. Andrieux (Diss. Univ. Paris 1929, pp. 43/72; C.A. **1929** 4893). — [62] J. M. Gomes, K. Uchida (U.S. 3902973 [1975]; C.A. **83** [1975] No. 185521). — [63] J. L. Andrieux (Rev. Met. **45** [1948] 49/59; J. Four Elec. **57** No. 3 [1948] 54). — [64] Z. Fisk, P. H. Schmidt, L. D. Longinotti (Mater. Res. Bull. **11** [1976] 1019/22). — [65] S. Aronson, E. Cisney, K. A. Gingerich (J. Am. Ceram. Soc. **50** [1967] 248/52).

[66] S. Aronson, A. Auskern (Thermodyn. Proc. Symp., Vienna 1965 [1966], Vol. 1, pp. 165/70). — [67] G. Allard (Compt. Rend. **189** [1929] 108/9). — [68] M. von Stackelberg (Z. Electrochem. **37** [1931] 542/5). — [69] M. von Stackelberg, F. Neumann (Z. Physik. Chem. B **19** [1932] 314/20). — [70] G. Allard (Bull. Soc. Chim. France **51** [1932] 1213/5).

[71] N. N. Zhuravlev, A. A. Stepanova, Yu. B. Paderno, G. V. Samsonov (Kristallografiya **6** [1961] 791/4; Soviet Phys.-Cryst. **6** [1962] 636/8; C.A. **56** [1962] 2965). — [72] R. Kiessling (Acta Chem. Scand. **4** [1950] 209/27). — [73] F. Bertaut, P. Blum (Compt. Rend. **234** [1952] 2621/3). — [74] O. I. Shulishova, I. A. Shcherbak (Neorgan. Materialy **3** [1967] 1495/7; Inorg. Materials [USSR] **3** [1967] 1304/6; C.A. **68** [1968] No. 44082). — [75] S. Aronson, A. Ingraham (J. Nucl. Mater. **24** [1967] 74/9).

[76] R. Kiessling (J. Electrochem. Soc. **98** [1951] 166/70). — [77] I. G. Barentseva, Yu. B. Paderno (Poroshkovaya Met. **1982** No. 21, pp. 83/7; Soviet Powder Met. Metal Ceram. **1982** No. 21, pp. 585/8; C.A. **97** [1982] No. 102177). — [78] I. G. Barentseva, Yu. B. Paderno (Vysokotemp. Boridy Silitsidy **1983** 8/12; C.A. **99** [1983] No. 185424). — [79] G. V. Samsonov, Yu. B. Paderno, E. E. Vainshtein (Izv. Sibirsk. Otd. Akad. Nauk SSSR Ser. Khim. Nauk **1964** No. 3, pp. 78/84; C.A. **63** [1965] 2401). — [80] Ya. I. Dutchak, Ya. I. Fedyshin, Yu. B. Paderno, D. I. Vadets (Izv. Vysshikh Uchebn. Zavedenii Fiz. **16** [1973] 154/6; C.A. **78** [1973] No. 129327).

[81] Ya. I. Dutchak, Ya. I. Fedyshin, Yu. B. Paderno, D. I. Vadets, V. V. Odintsov (Tezisy Dokl. 2nd Vses. Konf. Kristallokhim. Intermetal. Soedin., Lvov 1974, p. 149; C.A. **86** [1977] No. 10825). — [82] S. M. L'vov, V. F. Nemchenko, Yu. B. Paderno (Dokl. Akad. Nauk SSSR **149** [1963] 1371/2; C.A. **59** [1963] 4573). — [83] G. V. Samsonov (Poroshkovaya Met. **1963** No. 2, pp. 65/79; Soviet Powder Met. Metal Ceram. **1963** No. 2, pp. 139/50; C.A. **59** [1963] 278). — [84] S. Aronson, A. Ingraham (BNL-50023 [1966] 148; N.S.A. **21** [1967] No. 35701). — [85] Ya. I. Dutchak, Ya. I. Fedyshin, Yu. B. Paderno (Neorgan. Materialy **8** [1972] 2134/7; Inorg. Materials [USSR] **8** [1973] 1877/80; C.A. **78** [1973] No. 89548).

[86] G. V. Samsonov (Usp. Khim. **28** [1959] 189/217; C.A. **1959** 11074). — [87] S. Aronson, A. Ingraham (BNL-50082 [1967] 128; N.S.A. **22** [1968] No. 38510). — [88] J. M. Lafferty (Phys. Rev. [2] **79** [1950] 1012). — [89] G. V. Samsonov, E. E. Vainshtein, Yu. B. Paderno (Fiz. Metal. Metalloved. **13** [1962] 744/9; C.A. **57** [1962] 13257). — [90] Yu. B. Paderno, G. V. Samsonov (Dokl. Akad. Nauk SSSR **137** [1961] 646/7; Proc. Acad. Sci. USSR Phys. Chem. Sect. **136/141** [1961] 293/4; C.A. **57** [1962] 1674).

[91] J. N. Chazalviel, M. Campagna, G. K. Wertheim, P. H. Schmidt, D. L. Longinotti (Proc. 12th Rare Earth Res. Conf., Vail, Colo., 1976, Vol. 2, pp. 542/51; C.A. **85** [1976] No. 133690). — [92] L. N. Kugai, T. N. Nazarchuk (Zh. Analit. Khim. **16** [1961] 205/8; J. Anal. Chem. [USSR] **16** [1961] 213/6; C.A. **56** [1962] 12299). — [93] J. F. Cannon, H. T. Hall (Rare Earths Mod. Sci. Technol. **13** [1977/78] 219/24; C.A. **92** [1980] No. 14638). — [94] J. Etourneau, J. S. Kasper (J. Solid State Chem. **3** [1971] 101/11).

13.2   Ternary Compounds of Thorium and Boron

Ternary thorium oxides with boron (borates) such as ThB_2O_5 and the system ThO_2-B_2O_3 are dealt with in "Thorium" Erg.-Bd. C2, 1978, pp. 18/20, and are not described here.

Ternary thorium borides with Na or with rare earth metals are described in the following chapter. The related compounds with Fe, V, Mo, W, and Re are not described here in detail due to the Gmelin system of last position, but only briefly considered.

13.2.1   With Hydrogen

Ternary thorium hydrides with boron, such as $Th(BH_4)_4$, have already been described in "Thorium" Erg.-Bd. C 1, 1978, pp. 25/6. The description of $Th(BH_4)_4$ in this present chapter includes the newer literature. The derived compounds $LiTh(BH_4)_5$, $Li_2Th(BH_4)_6$, and $N(C_4H_9)_4Th(BH_4)_5$ are dealt with in the following sections.

13.2.1.1   Thorium Tetrahydroborate, $Th(BH_4)_4$

Preparation

Thorium tetrahydroborate was prepared from dry ThF_4, and $Al(BH_4)_3$, according to ThF_4 + 2 $Al(BH_4)_3$ = $Th(BH_4)_4$ + 2 $AlF_2(BH_4)$. The reaction was carried out with excess $Al(BH_4)_3$, condensed on the ThF_4 in an evacuated and sealed tube, at room temperature within several days. The excess $Al(BH_4)_3$ was then pumped off and the reaction tube heated to 150°C (up to the collar level) for several hours to disproportionate the $AlF_2(BH_4)$ formed to aluminium hydroborate, which was pumped off continuously, and the nonvolatile AlF_3. At this stage, the thorium hydroborate formed was sublimed and condensed above the heated zone as a white crystalline product [1], also cited in [2, 3], see also "Thorium" 1955, p. 296. $Th(BH_4)_4$ was also prepared in sufficient yield (50 to 78%) by disproportionation of the compounds $Li_2Th(BH_4)_6$ and $Li_2ThCl_2(BH_4)_4$ (see p. 140) at 150°C in a vacuum of 10^{-2} Torr. Under these conditions $Th(BH_4)_4$ sublimed [4].

Crystallographic and Optical Properties

An X-ray investigation showed $Th(BH_4)_4$ to be isomorphous with $U(BH_4)_4$ (see "Uran" Erg.-Bd. C1, 1977, pp. 73/9) [1].

The IR spectrum of Th(BH$_4$)$_4$ is quite similar to those of the compounds Zr(BH$_4$)$_4$ (see "Zirkonium" Erg.-Bd., 1958, pp. 359/60) and Hf(BH$_4$)$_4$ (see "Hafnium" Erg.-Bd., 1958, p. 22). As in Zr(BH$_4$)$_4$ [5], the Th(BH$_4$)$_4$ has a tridentate structure with a C$_{3v}$ symmetry. From this structure the following IR vibrations are assumed: one strong B-H stretching vibration at 2450 to 2600 cm^{-1} (A$_1$ symmetry), a strong bridge deformation vibration at 1150 to 1250 cm^{-1} (E symmetry), and two strong ThH$_3$B bridge vibrations at 2100 to 2200 cm^{-1} (A$_1$, E symmetry) [4, 6], see also [7]. The measured vibration frequencies are summarized in Table 67.

^{11}B NMR data for Th(BH$_4$)$_4$ (dissolved in ether) are δ ^{11}B = +8.0 ppm (related to BF$_3 \cdot$ (C$_2$H$_5$)$_2$O = 0 ppm) and J$_{BH}$ = 86.5 Hz [4].

Table 67
IR Frequencies of Th(BH$_4$)$_4$ and Th(BD$_4$)$_4$ (in cm^{-1}).
s = strong, m = medium intensity, sh = shoulder.

vibration	Th(BH$_4$)$_4$ [7]	Th(BH$_4$)$_4$ [4]	Th(BD$_4$)$_4$ [4]
vBH (BD)	2555 (m)	2530	1880
terminal	2545 (s)		
B-H vibration	2505 (m)		
	2440 (m)		
vThH$_3$B (ThD$_3$B)	2285 (s)	2270	1700
bridge	2235 (s)	2200	1620
B-H vibration	2200 (s)	2100	1550
	2118 (s)		
δBH (BD)	1260 (sh)		
deformation	1188 (s)	1165	800
vibration	1116 (s)		
	1080 (s)		
	1015 (s)		
δThB$_4$	462 (s)		
vibration	450 (s)	450	490

Thermal Properties

These are dealt with in "Thorium" 1955, p. 297, according to [1], and also cited in [2, 3].

The vapor pressure follows the equation log p (in Torr) = $-$A/T + B, where A = 2844 and B = 10719 [1]. No thermal effects were observed on DTA measurements in the range $-$180 to 55°C; at 158°C, one exothermic effect was found due to irreversible decomposition [7].

Melting point is 200 to 205°C [4].

Chemical Reactions

See "Thorium" 1955, p. 297. From DTA measurements thermal decomposition was observed at 158°C [7]. Th(BH$_4$)$_4$ is normally handled under inert conditions, but it is indefinitely stable at room temperature. Th(BH$_4$)$_4$ is decomposed in water at room temperature forming

References for 13.2.1.1 on p. 140

thorium borate and hydrogen. The reaction is rather vigorous; small amounts may be dropped into water without exploding [1], see also [2]. See "Thorium" 1955, p. 297.

References for 13.2.1.1:

[1] H. R. Hoekstra, J. J. Katz (J. Am. Chem. Soc. **71** [1949] 2488/92). − [2] L. I. Katzin (Natl. Nucl. Energy Ser. Div. IV B **14** [1954] 66/102). − [3] L. I. Katzin (Proc. Intern. Conf. Peaceful Uses At. Energy, Geneva 1955, Vol. 7, pp. 401/13, Paper P/734). − [4] M. Ehemann, H. Nöth (Z. Anorg. Allgem. Chem. **386** [1971] 87/101). − [5] P. H. Bird, M. R. Churchill (Chem. Commun. **1967** 403).

[6] T. J. Marks, W. J. Kenelly, J. R. Kolb, L. A. Shimp (Inorg. Chem. **11** [1972] 2540/6). − [7] V. V. Volkov, K. G. Myakishev, Z. A. Grankina (Zh. Neorgan. Khim. **15** [1970] 2861/2; Russ. J. Inorg. Chem. **15** [1970] 1490/1; N.S.A. **25** [1971] No. 8263).

13.2.1.2 LiTh(BH$_4$)$_5$

The reaction of LiBH$_4$ with Th(BH$_4$)$_4$ in diethyl ether at room temperature leads to the formation of LiTh(BH$_4$)$_5 \cdot$ n(C$_2$H$_5$)$_2$O, from which at 50°C in vacuum pure LiTh(BH$_4$)$_5$ was obtained. LiTh(BH$_4$)$_5$ decomposes at 149°C, changing its color from white to brown and evolving a gas. From IR and ^{11}B NMR spectra the BH$_4$ groups are bonded to the Th atoms via 3 hydrogen bridges. The characteristic IR frequencies are 2430 cm^{-1} for the terminal B-H vibrations (vBH), 2270 cm^{-1} and 2215 cm^{-1} for the bridge B-H vibrations (vThH$_3$B), and 1160 cm^{-1} for the deformation vibrations (δBH). ^{11}B NMR data for LiTh(BH$_4$)$_5$ dissolved in ether are δ ^{11}B = +13.4 ppm (related to BF$_3 \cdot$ (C$_2$H$_5$)$_2$O = 0 ppm), and J$_{BH}$ = 84.5 Hz or 79.5 Hz if dissolved in tetrahydrofuran [1].

13.2.1.3 Li$_2$Th(BH$_4$)$_6$. Li$_2$ThCl$_2$(BH$_4$)$_4$

The reaction of LiBH$_4$ with Th(BH$_4$)$_4$ in diethyl ether at room temperature leads to the formation of Li$_2$Th(BH$_4$)$_6 \cdot$ m(C$_2$H$_5$)$_2$O, from which at 50°C in vacuum pure Li$_2$Th(BH$_4$)$_6$ was obtained. Li$_2$Th(BH$_4$)$_6$ could not be prepared from the reaction of LiBH$_4$ and ThCl$_4$, but a dichloro compound was recovered from the solution in ether according to 4 LiBH$_4$ + ThCl$_4$ = Li$_2$ThCl$_2$(BH$_4$)$_4$ + 2 LiCl. This compound decomposes to Th(BH$_4$)$_4$ at 150°C in a vacuum of 10^{-2} Torr. The white crystalline Li$_2$Th(BH$_4$)$_6$ decomposes at temperatures above 150°C.

^{11}B NMR data for Li$_2$Th(BH$_4$)$_6$ dissolved in ether are δ ^{11}B = +18.9 ppm (related to BF$_3 \cdot$ (C$_2$H$_5$)$_2$O = 0 ppm) and J$_{BH}$ = 84.5 Hz. [1].

13.2.1.4 N(C$_4$H$_9$)$_4$Th(BH$_4$)$_5$

The compound has been prepared from a suspension of Th(BH$_4$)$_4$ in benzene by addition of tributyl ammonium borate dissolved in benzene. Recovery of the precipitate after 1 h of stirring and drying in a high vacuum for about 30 min leads to a white crystalline solid (melting point 112°C) of composition N(C$_4$H$_9$)$_4$Th(BH$_4$)$_5$. The BH$_4$ groups are bonded to the Th atoms via three hydrogen bridges, as IR and ^{11}B NMR spectra showed. The characteristic IR frequencies are $2440\ cm^{-1}$ for the terminal B-H vibrations (vBH), $2240\ cm^{-1}$, $2200\ cm^{-1}$, $2175\ cm^{-1}$ for the bridge B-H vibrations (vThH$_3$B), and $1175\ cm^{-1}$ for the deformation vibrations (vBH). ^{11}B NMR data for $N(C_4H_9)_4Th(BH_4)_5$ dissolved in tetrahydrofuran are $\delta\ ^{11}B = +13.7$ ppm (related to $BF_3 \cdot (C_2H_5)_2O = 0$ ppm) and $J_{BH} = 82$ Hz [1].

Reference for 13.2.1.2 to 13.2.1.4:

[1] M. Ehemann, H. Nöth (Z. Anorg. Allgem. Chem. **386** [1971] 87/101).

13.2.2   With Sodium

$Th_xNa_{1-x}B_6$ solid solutions were prepared within the experimental limits of $0.23 \leqslant x \leqslant 1$. Lattice parameters of the cubic crystals are a = 4.127 Å for $Th_{0.56}Na_{0.44}B_6$, a = 4.142 Å for $Th_{0.37}Na_{0.63}B_6$, and a = 4.151 Å for $Th_{0.23}Na_{0.77}B_6$ [1].

Reference for 13.2.2:

[1] F. Bertaut, P. Blum (Compt. Rend. **234** [1952] 2621/3).

13.2.3   With Rare Earth Elements

13.2.3.1   Phase Relationships

ThB$_4$ und CeB$_4$ form a complete series of solid solutions at 1600 °C with tetragonal structure [1, 2]. Complete series of solid solutions with cubic structure are formed with ThB$_6$ and LnB$_6$ (Ln = Ce, Y, Sm) [1, 3, 4]. A tentative phase diagram of the system cerium-thorium-boron (at 1600 °C) is given for the boride region in **Fig. 50**.

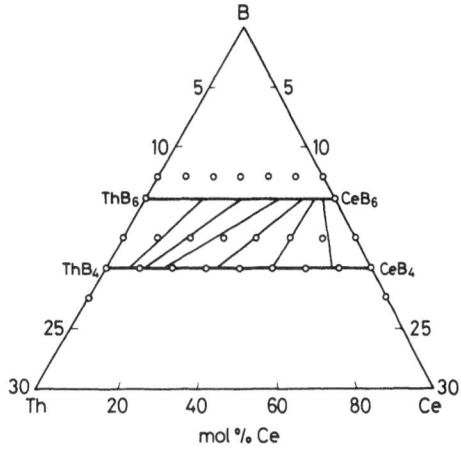

Fig. 50. Tentative phase diagram for the system Ce-Th-B at 1600 °C [2].

   References for 13.2.3 on p. 146

13.2.3.2 $Ce_{1-x}Th_xB_4$ Solid Solutions

$Ce_{1-x}Th_xB_4$ compounds are prepared from powdered thorium, cerium hydride, and boron, mixed, pressed into pellets (6 t/cm²), and placed in tungsten crucibles. The reaction was carried out at 1600°C within 10 h in a vacuum of 10^{-4} to 10^{-5} Torr. The reaction product was re-ground, re-pressed, and heated again for homogenization. The lattice parameters of the tetragonal $Ce_{1-x}Th_xB_4$ compounds (at 1600°C) show a slight positive deviation from the Végard law (see **Fig. 51**). The interaction parameter, a, between ThB_4 and CeB_4 was calculated from the tie line position at 1600°C (see Fig. 51) due to $ThB_4 + CeB_6 \rightleftharpoons CeB_4 + ThB_6$ to be $a_{ThB_4, CeB_4} = +4500$ cal/mol with the assumption of a regular behavior of the components due to the free energy equation $\Delta G = \Delta G° + RT \cdot \ln x + a(1-x)^2$. From this, separation of the solid solutions into the components should occur below 1000°C [1, 2].

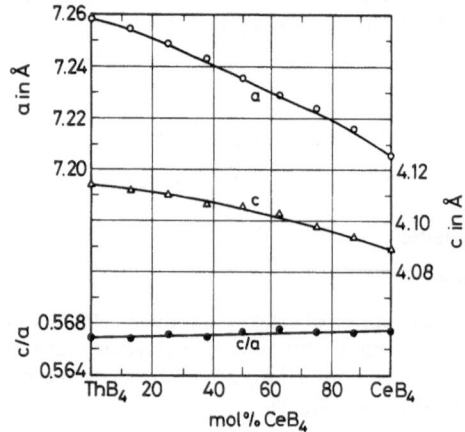

Fig. 51. Lattice parameters of the $Ce_{1-x}Th_xB_4$ compounds [1].

13.2.3.3 $Ln_{1-x}Th_xB_6$ Solid Solutions (Ln = Ce, Y, Sm)

Preparation

$Ce_{1-x}Th_xB_6$ compounds were prepared from powdered thorium, cerium hydride, and boron, pressed into pellets (6 t/cm²) and placed in tungsten crucibles. The reaction was carried out at 1600°C within 10h in a vacuum of 10^{-4} to 10^{-5} Torr. The reaction product was re-ground, re-pressed, and heated again for homogenization [1, 2], see also [5]. $Ce_{1-x}Th_xB_6$ compounds were also obtained by electrolytical reduction of ThO_2 and CeO_2 fused in B_2O_3 (105 g), MgO (40 g), and MgF_2 (70 g). The electrolysis was carried out at 1150°C within 2 h at 8 V and 25 A with a graphite anode and a graphite crucible as cathode [6], see also [7]. $Y_{1-x}Th_xB_6$ and $Sm_{1-x}Th_xB_6$ compounds were prepared by borothermic reduction of the corresponding oxides (Y_2O_3 or Sm_2O_3, ThO_2). The mixed powders were pressed into pellets and heated in an induction furnace at about 1700°C in a vacuum of 10^{-5} Torr using ZrB_2 crucibles. The reaction product was re-ground and heated again for homogenization [3, 4].

Crystallographic Properties

The lattice parameters of the cubic $Ce_{1-x}Th_xB_6$ solid solutions show only a very slight deviation from the Végard law (see **Fig. 52**) [1], whereas strong deviations from the Végard law were observed for the solid solutions of $Y_{1-x}Th_xB_6$ near YB_6 (see Table 68) [3] and of $Sm_{1-x}Th_xB_6$ (see **Fig. 53**) [4].

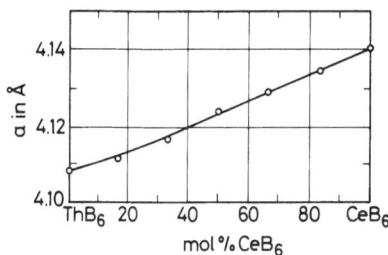

Fig. 52. Lattice parameters for $Ce_{1-x}Th_xB_6$ compounds [1].

Table 68
Lattice Parameters of $Y_{1-x}Th_xB_6$ Compounds [3].

x in $Y_{1-x}Th_xB_6$	a in Å ($\pm$ 0.001 Å)	x in $Y_{1-x}Th_xB_6$	a in Å ($\pm$ 0.001 Å)
0	4.1450	0.59	4.1051
0.12	4.1009	0.65	4.1057
0.30	4.1023	0.72	4.1062
		1	4.1099

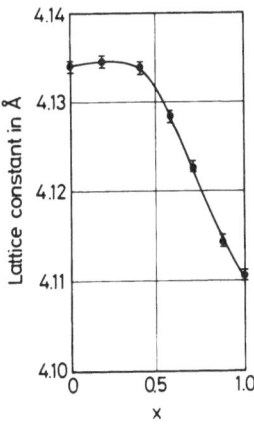

Fig. 53. Lattice parameters for $Sm_{1-x}Th_xB_6$ compounds [4].

References for 13.2.3 on p. 146

Thermal Stability

The interaction parameter, a, between ThB_6 and CeB_6 was calculated from the tie line position at 1600°C (see Fig. 50, p. 141) due to $ThB_6 + CeB_4 \rightleftharpoons CeB_6 + ThB_4$ to be $a_{ThB_6, CeB_6} = +3800$ cal/mol with the assumption of a regular behavior of the components according to the free energy equation $\Delta G = \Delta G° + RT \cdot \ln x + a(1-x)^2$. From this, separation of the solid solutions into the components should occur below 1000°C [1, 2].

Thermionic Emission

The current density, I, of the thermionic emission follows the equation $\log I/T^2 = \log A -11600 \, \varphi/2.303 \, T$ with the work function $\varphi = 2.92$ eV (for ThB_6), 2.59 eV (for CeB_6) and the constant $A = 0.5 \, A \cdot cm^{-2} \cdot K^{-2}$ (for ThB_6) and $3.6 \, A \cdot cm^{-2} \cdot K^{-2}$ (for CeB_6). The mixed boride $Ce_{0.5}Th_{0.5}B_6$ gave an emission current just between the emission currents of the pure components (see **Fig. 54**) [5], also cited in [7].

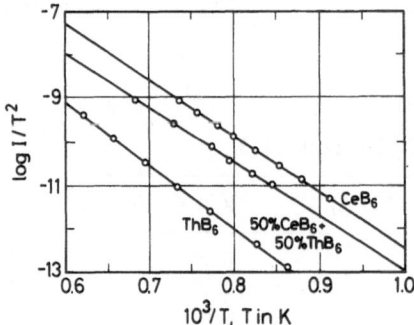

Fig. 54. Richardson plots for CeB_6, $Ce_{0.5}Th_{0.5}B_6$, and ThB_6 [5].

X-ray Absorption Spectrum

X-ray absorption spectra have been measured on $Sm_{1-x}Th_xB_6$ samples with 1.72 GeV electrons at 300 K. The spectra show two Sm peaks, separated by 7 eV, corresponding to the valence states of Sm^{2+} and Sm^{3+}. As a result, the ratio of the relative intensities $I(Sm^{2+}):I(Sm^{3+})$ increases with increasing values of x in $Sm_{1-x}Th_xB_6$ (see Fig. 57). Thus, the $Sm_{1-x}Th_xB_6$ solid solutions are compounds with mixed valence state, $(Sm^{2+}_{1-y}Sm^{3+}_y)_{1-x}Th_xB_6$ [4].

Superconductivity

The superconductivity of $Y_{1-x}Th_xB_6$ samples was measured by a flux expulsion method. The superconducting transition temperatures decrease from slightly below 6 K (for YB_6) to less than 1.28 K (for ThB_6) [8] with increasing values of x (**Fig. 55**) [3].

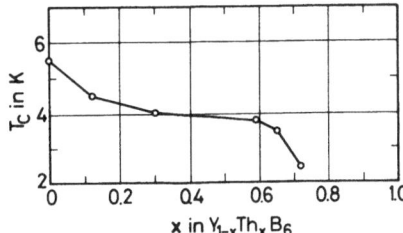

Fig. 55. Superconducting transition temperatures, T_c, for $Y_{1-x}Th_xB_6$ compounds [3].

Magnetic Susceptibility

The magnetic susceptibility of $Ce_{1-x}Th_xB_6$ compounds was measured at 4.2 to 300 K (see **Fig. 56**). At room temperature the values are between those of the free Sm^{2+} and Sm^{3+} cations: $\chi = z\,\chi_{Sm^{3+}} + (1-z)\,\chi_{Sm^{2+}}$ [4]. The average samarium valences obtained are shown in **Fig. 57**.

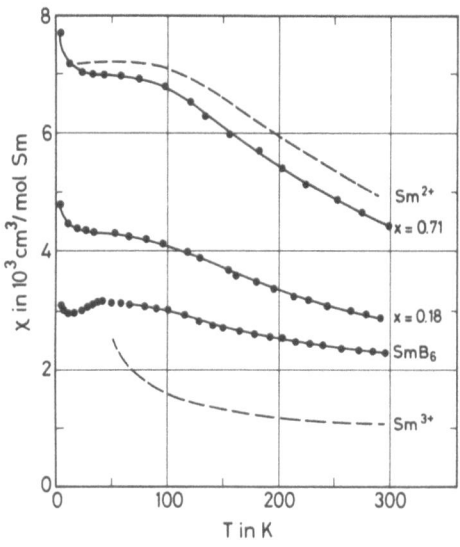

Fig. 56. Magnetic susceptibility for $Sm_{1-x}Th_xB_6$ compounds [4].

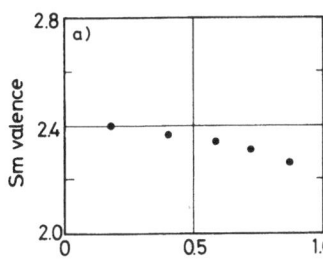

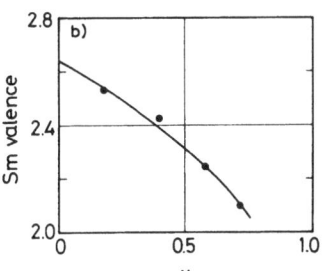

Fig. 57. Average samarium valence in $Sm_{1-x}Th_xB_6$ compounds [4]; a) from X-ray absorption measurements; b) from magnetic susceptibility measurements.

References for 13.2.3 on p. 146

The magnetic susceptibility of diamagnetic $Y_{0.6}Th_{0.4}B_6$ was measured at 1.5 to 300 K. The measured value of $\chi_g = -0.12 \cdot 10^{-6}$ cm³/g remains down to 6 K. In the $Y_{1-x}Th_xB_6$ series no magnetic moment is involved [3].

References for 13.2.3:

[1] P. Stecher, F. Benesovsky, H. Nowotny (Planseeber. Pulvermet. **13** [1965] 37/46; C.A. **63** [1965] 9582). — [2] F. Benesovsky, P. Stecher, H. Nowotny, W. Rieger (Colloq. Intern. Centre Natl. Rech. Sci. [Paris] No. 157 [1967] 419/30). — [3] K. Hiebl, M. J. Sienko (Inorg. Chem. **19** [1980] 2179/80). — [4] J. M. Tarascon, Y. Isikawa, B. Chevalier, J. Etourneau, P. Hagenmuller, M. Kasaya (J. Phys. [Paris] **41** [1980] 1135/40). — [5] J. M. Lafferty (J. Appl. Phys. **22** [1951] 299/309).

[6] L. Andrieux (Compt. Rend. **194** [1932] 720/2). — [7] S. Peterson (ORNL-4503 Vol. 4 [1971] 1/13; C.A. **78** [1973] No. 139584). — [8] O. I. Shulishova, I. A. Shcherbak (Izv. Akad. Nauk SSSR Neorgan. Materialy **3** [1967] 1495/7; Inorg. Materials [USSR] **3** [1967] 1304/6; C.A. **68** [1968] No. 44082).

13.2.4   With Other Metals

$Th_2Fe_{14}B$

$Th_2Fe_{14}B$ samples were prepared from the elements by arc-melting under argon of high purity. The arc-melted samples were then wrapped in tantalum foil, sealed in evacuated quartz tubes, and annealed at 900 °C for 3 weeks for homogenization. The grains in the samples showed a substantial texture. $Th_2Fe_{14}B$ crystallizes with tetragonal structure ($Nd_2Fe_{14}B$ type); the proposed space group is $P4_2/mm\text{-}D_{2h}^{14}$ (No. 136). Magnetization experiments performed on $Th_2Fe_{14}B$ samples were carried out at 4.2 to 1000 K in magnetic fields up to 1440 kA/m using a modified Faraday method. The field dependence of the magnetization at room temperature is shown in **Fig. 58**, measured in the direction parallel, $\sigma_{\parallel}$, and perpendicular, $\sigma_{\perp}$, to the alignment field. From this, an anisotropy field, H_A, is deduced in the order of $2.4 \cdot 10^3$ kA/m (30 kOe). Its Curie temperature is given to be 480 K. Hyperfine field parameters of $Th_2Fe_{14}B$ were obtained from ^{57}Fe Mössbauer spectra. From the Mössbauer spectra it was deduced that different crystallographic iron sites give rise to different values of the iron moments. The average iron moment in $Th_2Fe_{14}B$ is only slightly lower than that in $Nd_2Fe_{14}B$ (2.1 μ_B per iron atom) [1].

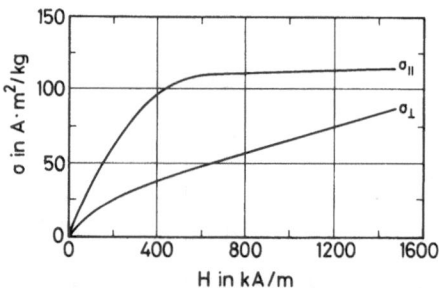

Fig. 58. Field dependence of the magnetization in $Th_2Fe_{14}B$ at room temperature measured on an aligned piece of polycrystalline material with the external field applied parallel, $\sigma_{\parallel}$, and perpendicular, $\sigma_{\perp}$, to the alignment field [1].

Ternary Thorium Compounds with ThMoB₄ Structure

Ternary compounds of thorium with orthorhombic structure (ThMoB$_4$ type) have been found to exist for V, Mo, W, and Re. Analogous compounds with niobium and chromium could not be prepared [2]. A phase diagram for the system thorium-tungsten-boron based on X-ray diffraction measurements is given in [3].

ThMB$_4$ (M = V, Mo, W, Re) were prepared from the elements, mixed and pressed into pellets, by arc-melting under argon of high purity. The samples were re-melted for homogenization [2]. The ternary compounds were also obtained by sintering of the mixed and pressed metal powders in high-vacuum furnaces at 1800°C for ½ h (10^{-4} to 10^{-5} Torr) (ThWB$_4$ [3]) or at 1750°C for 1 h (10^{-6} Torr) with the samples placed on a special ThO$_2$ disk [3] or on molybdenum or tungsten sheets [2]. Small fragments of single crystals could be recovered from the arc-melted samples [2]. With both preparative techniques well crystallized samples were obtained, but small amounts of ThO$_2$ as an impurity could not be avoided [2, 3].

ThWB$_4$ was determined from X-ray diffraction pattern to crystallize with monoclinic structure, Z = 2, with an assumed space group of P2-C$_2^1$ (No.3), P2/m-C$_{2h}^1$ (No. 10), or Pm-C$_3^1$ (No. 6), and lattice constants of a = 12.25, b = 3.75, c = 6.14 Å, β = 104.1°. The X-ray density was calculated to be 5.6 g/cm^3 as compared to a measured density of 5.8 g/cm^3 [3].

A re-investigation of the crystal structure of the ThMB$_4$ compounds (M = V, Mo, W, Re) performed on small single crystals of ThMoB$_4$ and ThWB$_4$ using Weissenberg patterns resulted in a base-centered orthorhombic unit cell with Z = 4, the space group is Cmmm-D$_{2h}^{11}$ (No. 65) [2]. The measured lattice parameters and the cell volumes are summarized in Table 69.

Table 69
Lattice Parameters and Cell Volumes of Orthorhombic ThMB$_4$ Compounds (M = V, Mo, W, Re) [2].

compound	a in Å	b in Å	c in Å	V in Å^3
ThVB$_4$	7.453	9.623	3.682	264.1
ThMoB$_4$	7.481	9.658	3.771	272.5
ThWB$_4$	7.487	9.681	3.739	271.0
ThReB$_4$	7.436	9.606	3.715	265.4

The crystals of the ThMoB$_4$ type compounds are built up in a layer structure consisting of a two-dimensional network of boron atoms forming rings with five or seven boron atoms and layers of thorium and molybdenum atoms [2]. The atomic parameters and the interatomic distances for the compound ThMoB$_4$ are given in the original paper.

References for 13.2.4:

[1] K. H. J. Buschow, H. M. van Noort, D. B. de Mooij (J. Less-Common Metals **109** [1985] 79/91]. − [2] P. Rogl, H. Nowotny (Monatsh. Chem. **105** [1974] 1082/98). − [3] D. T. Pitman, D. K. Das (J. Electrochem. Soc. **107** [1960] 763/6).

Table of Conversion Factors

Following the notation in Landolt-Börnstein [7], values that have been fixed by convention are indicated by a bold-face last digit. The conversion factor between calorie and Joule that is given here is based on the thermochemical calorie, cal_{thch}, and is defined as 4.1840 J/cal. However, for the conversion of the "Internationale Tafelkalorie", cal_{IT}, into Joule, the factor 4.1868 J/cal is to be used [1, p. 147]. For the conversion factor for the British thermal unit, the Steam Table Btu, BTU_{ST}, is used [1, p. 95].

Force	N	dyn	kp
1 N (Newton)	1	10^5	0.1019716
1 dyn	10^{-5}	1	1.019716×10^{-6}
1 kp	9.80665	9.80665×10^5	1

Pressure	Pa	bar	kp/m²	at	atm	Torr	lb/in²
1 Pa (Pascal) = 1N/m²	1	10^{-5}	1.019716×10^{-1}	1.019716×10^{-5}	0.986923×10^{-5}	0.750062×10^{-2}	145.0378×10^{-6}
1 bar = 10^6 dyn/cm²	10^5	1	10.19716×10^3	1.019716	0.986923	750.062	14.50378
1 kp/m² = 1mm H_2O	9.80665	0.980665×10^{-4}	1	10^{-4}	0.967841×10^{-4}	0.735559×10^{-1}	1.422335×10^{-3}
1 at = 1 kp/cm²	0.980665×10^5	0.980665	10^4	1	0.967841	735.559	14.22335
1 atm = 760 Torr	1.01325×10^5	1.01325	1.033227×10^4	1.033227	1	760	14.69595
1 Torr = 1mm Hg	133.3224	1.333224×10^{-3}	13.59510	1.359510×10^{-3}	1.315789×10^{-3}	1	19.33678×10^{-3}
1 lb/in² = 1 psi	6.89476×10^3	68.9476×10^{-3}	703.069	70.3069×10^{-3}	68.0460×10^{-3}	51.7149	1

Work, Energy, Heat	J	kWh	kcal	Btu	MeV
1 J (Joule) = 1 Ws = 1 Nm = 10^7 erg	1	2.778×10^{-7}	2.39006×10^{-4}	9.4781×10^{-4}	6.242×10^{12}
1 kWh	3.6×10^6	1	860.4	3412.14	2.247×10^{19}
1 kcal	4184.0	1.1622×10^{-3}	1	3.96566	2.6117×10^{16}
1 Btu (British thermal unit)	1055.06	2.93071×10^{-4}	0.25164	1	6.5858×10^{15}
1 MeV	1.602×10^{-13}	4.450×10^{-20}	3.8289×10^{-17}	1.51840×10^{-16}	1

1 eV ≙ 23.0578 kcal/mol = 96.473 kJ/mol

Power	kW	PS	kp m/s	kcal/s
1 kW = 10^{10} erg/s	1	1.35962	101.972	0.239006
1 PS	0.73550	1	75	0.17579
1 kp m/s	9.80665×10^{-3}	0.01333	1	2.34384×10^{-3}
1 kcal/s	4.1840	5.6886	426.650	1

References:

[1] A. Sacklowski, Die neuen SI-Einheiten, Goldmann, München 1979. (Conversion tables in an appendix.)
[2] International Union of Pure and Applied Chemistry, Manual of Symbols and Terminology for Physicochemical Quantities and Units, Pergamon, London 1979; Pure Appl. Chem. 51 [1979] 1/41.
[3] The International System of Units (SI), National Bureau of Standards Spec. Publ. 330 [1972].
[4] H. Ebert, Physikalisches Taschenbuch, 5th Ed., Vieweg, Wiesbaden 1976.
[5] Kraftwerk Union Information, Technical and Economic Data on Power Engineering, Mülheim/Ruhr 1978.
[6] E. Padelt, H. Laporte, Einheiten und Größenarten der Naturwissenschaften, 3rd Ed., VEB Fachbuchverlag, Leipzig 1976.
[7] Landolt-Börnstein, 6th Ed., Vol. II, Pt. 1, 1971, pp. 1/14.
[8] ISO Standards Handbook 2, Units of Measurement, 2nd Ed., Geneva 1982.

Key to the Gmelin System
of Elements and Compounds

System Number	Symbol	Element
1		Noble Gases
2	H	Hydrogen
3	O	Oxygen
4	N	Nitrogen
5	F	Fluorine
6	**Cl**	**Chlorine**
7	Br	Bromine
8	I	Iodine
	At	Astatine
9	S	Sulfur
10	Se	Selenium
11	Te	Tellurium
12	Po	Polonium
13	B	Boron
14	C	Carbon
15	Si	Silicon
16	P	Phosphorus
17	As	Arsenic
18	Sb	Antimony
19	Bi	Bismuth
20	Li	Lithium
21	Na	Sodium
22	K	Potassium
23	NH_4	Ammonium
24	Rb	Rubidium
25	Cs	Caesium
	Fr	Francium
26	Be	Beryllium
27	Mg	Magnesium
28	Ca	Calcium
29	Sr	Strontium
30	Ba	Barium
31	Ra	Radium
32	**Zn**	**Zinc**
33	Cd	Cadmium
34	Hg	Mercury
35	Al	Aluminium
36	Ga	Gallium

System Number	Symbol	Element
37	In	Indium
38	Tl	Thallium
39	Sc, Y La—Lu	Rare Earth Elements
40	Ac	Actinium
41	Ti	Titanium
42	Zr	Zirconium
43	Hf	Hafnium
44	Th	Thorium
45	Ge	Germanium
46	Sn	Tin
47	Pb	Lead
48	V	Vanadium
49	Nb	Niobium
50	Ta	Tantalum
51	Pa	Protactinium
52	**Cr**	**Chromium**
53	Mo	Molybdenum
54	W	Tungsten
55	U	Uranium
56	Mn	Manganese
57	Ni	Nickel
58	Co	Cobalt
59	Fe	Iron
60	Cu	Copper
61	Ag	Silver
62	Au	Gold
63	Ru	Ruthenium
64	Rh	Rhodium
65	Pd	Palladium
66	Os	Osmium
67	Ir	Iridium
68	Pt	Platinum
69	Tc	Technetium[1]
70	Re	Rhenium
71	Np,Pu . . .	Transuranium Elements

HCl

$CrCl_2$

$ZnCrO_4$

$ZnCl_2$

Material presented under each Gmelin System Number includes all information concerning the element(s) listed for that number plus the compounds with elements of lower System Number.

For example, zinc (System Number 32) as well as all zinc compounds with elements numbered from 1 to 31 are classified under number 32.

[1] A Gmelin volume titled "Masurium" was published with this System Number in 1941.

A Periodic Table of the Elements with the Gmelin System Numbers is given on the Inside Front Cover